Biotechnology
Quality Assurance
and Validation

■ ■ ■

Drug Manufacturing
Technology Series

Volume 4

Edited by
Kenneth E. Avis
Carmen M. Wagner
Vincent L. Wu

CRC Press
Taylor & Francis Group
Boca Raton London New York

CRC Press is an imprint of the
Taylor & Francis Group, an **informa** business

CRC Press
Taylor & Francis Group
6000 Broken Sound Parkway NW, Suite 300
Boca Raton, FL 33487-2742

First issued in paperback 2019

© 1999 by Taylor & Francis Group, LLC
CRC Press is an imprint of Taylor & Francis Group, an Informa business

No claim to original U.S. Government works

ISBN-13: 978-1-57491-089-6 (hbk)
ISBN-13: 978-0-367-40025-5 (pbk)

Library of Congress Card Number 98-40953

Library of Congress Cataloging-in-Publication Data

Biotechnology : quality assurance and validation / Kenneth E. Avis, Carmen M. Wagner, Vincent L. Wu, editors.
    p. cm. - (Drug manufacturing technology series : v. 4.)
Includes bibliographical references and index.
ISBN 1-57491-089-2
1. Pharmaceutical biotechnology—Quality control. I. Avis, Kenneth E., 1918-. II. Wagner, Carmen M. III. Wu, Vincent L. IV. Series.

53

Visit the Taylor & Francis Web site at
http://www.taylorandfrancis.com

and the CRC Press Web site at
http://www.crcpress.com

CONTENTS

3. Quality Control and Quality Assurance Issues in Biopharmaceutical Processing   33

Gary D. Christiansen

5. Validation of Biopharmaceutical Processes   111

Howard L. Levine
Francisco J. Castillo

FOREWORD

This fourth volume in the *Drug Manufacturing Technology Series* of applied reference books returns to the biotechnology area, the subject area of Volume 2. As identified previously, this series of books is intended to provide practical details of how to perform selected tasks, with only sufficient theory or basic principles included to provide the foundation for a presentation of significant applied issues. Validation requirements are sometimes included within the purview of quality assurance but, if not, they are closely allied. Therefore, these topics are included together in this volume, "Biotechnology: Quality Assurance and Validation."

Practical details of how to perform selected tasks have been given for a range of issues, from storage control of viable cells to virus removal by filtration. It should not be assumed that either the topics of quality assurance or validation have been exhausted in this relatively small volume, but selected topics have been addressed. Included are presentations of inventory systems, quality assurance and quality control issues, a field investigator's view of manufacturing issues, validation of processes, cleaning, and virus removal by filtration.

As in the previous volumes, an extensive index has been prepared to enhance the inherent utility of the book. In addition, the Key Concept Index has been continued as a means of integrating this with other volumes in the series. In order for the latter tool to be useful and not become too complex, two Key Concept Indexes are being prepared, one incorporating topics related to the manufacture of Sterile Dosage Forms and one to Nonsterile Dosage Forms. The topics in this volume will be incorporated into the Key Concept Index for the manufacture of Sterile Dosage Forms.

I hope that you will have as much pleasure and benefit from using this book as I have had in bringing it together. There is immeasurable pleasure in bringing together a book such as this one, because I believe that the authors have provided information that will meet a critical need in the field of biotechnology. But, that pleasure can only be brought to fruition if you find it beneficial in the furtherance of your work. I sincerely hope that this will be the outcome.

Kenneth E. Avis, D.Sc.
Coordinating Editor

AUTHOR BIOGRAPHIES

KENNETH E. AVIS

Dr. Kenneth Avis is a distinguished leader in pharmaceutical science. He is past president of the PDA and has served the U.S. Pharmacopeia and U.S. FDA in numerous advisory capacities. Dr. Avis has consulted with more than 50 pharmaceutical companies, universities, hospitals, and governmental agencies; published over 30 peer-reviewed research papers; and is currently serving as Editor for the *Drug Manufacturing Technology Series* for Interpharm Press while maintaining his affiliation with the University of Tennessee, Memphis, as Emeritus Professor in Pharmaceutics.

GREGORY BOBROWICZ

At the FDA, Gregory Bobrowicz was a charter member of the landmark Pacific Regional Biotechnology Team. For 6-1/2 years, he conducted and managed preapproval and regulatory inspections of biologics, drugs, and devices; additionally, he drafted inspection guidance and compliance policy for biotechnology. Currently, he is a Senior Quality Systems Associate for Quintiles, where he assists manufacturers with gap analysis audits and corrective action project plans.

FRANCISCO J. CASTILLO

Dr. Francisco Castillo is currently Scientific Director, Head of Fermentation and Cell Culture Development for Berlex Biosciences. His responsibilities at Berlex include process development for the production of recombinant proteins and, more recently, vectors for Gene Therapy. Before joining Berlex in 1995, Dr. Castillo spent 10 years at Xoma Corporation as Director of Cell Culture and Fermentation Development. He was also a Visiting Scholar at U.C. Berkeley from 1982 to 1984, and Head of the Fermentation Laboratory at the Venezuelan Institute of Scientific Research in Caracas from 1976 to 1982.

GARY D. CHRISTIANSEN

For the past 6 years, Gary D. Christiansen, PhD, has been is a consultant for validation and compliance programs with Validation & Compliance Services (Gary D. Christiansen & Associates) in San Diego, Calif. Previous experience includes 12 years in positions that included Corporate Validations, Director of Therapeutic Quality Control and Quality Assurance, and Director of Clinical Product Production at Hybritech, Incorporated, San Diego, Calif.—a division of Eli Lilly. Prior to this, an additional 8 years of experience were gained as Director of Research and Process Development at Bio-Reagents—a division of Ortho Diagnostics, Irvine, Calif. Responsibilities have included establishing and maintaining quality functions, such as facility, equipment and process validation; GMP systems and training; and quality control laboratories.

HAZEL ARANHA-CREADO

Dr. Hazel Aranha-Creado is a Senior Staff Scientist at Pall Corporation with over 20 years of experience in Applied and Environmental Microbiology. She has an MS in Virology and a PhD in Environmental Microbiology. She has been with Pall Corporation for the last 5 years, where her responsibilities include technical assistance in the pharmaceutical, biomedical, and healthcare areas, with special emphasis on virus removal applications.

HOWARD L. LEVINE

Dr. Howard Levine is president of BioProcess Technology Consultants, a consulting firm specializing in biopharmaceutical process development, manufacturing, and engineering. Dr. Levine has over 20 years of experience in the biopharmaceutical industry and was previously vice president of Manufacturing Operations at Repligen Corporation. He has also worked in process development and manufacturing for Amgen, Inc., Genentech, Inc., and Xoma Corporation.

ROBERT W. O'BRIEN

Robert O'Brien is manager of validation for Biopure Corporation, a biopharmaceutical company focused on the development and manufacture of Oxygen Therapeutics (an alternative means to oxygenate tissues). Robert has been in the biopharmaceutical industry for several years, serving in various positions leading to his present position of two plus years. He is a 1982 graduate of Northeastern University College of Engineering with a Bachelor of Science degree in Industrial Engineering. Robert is a member of both the ISPE (Boston Chapter) and the PDA.

FRANK P. SIMIONE

Frank Simione is currently Director of Professional Services and Safety Officer at American Type Culture Collection (ATCC). Frank was Director of Operations at ATCC from 1988 to 1996 with overall responsibility for the inventory and distribution of ATCC cultures and related items. His primary interest is low-temperature preservation and maintenance of microorganisms and cells, and he has published and lectured widely on low-temperature methodologies for preservation and storage.

JON R. VOSS

Jon Voss is the president of Kemper-Masterson Systems, a division of KMI/PAREXEL formed to create validation and compliance software tools. In addition to his duties with KMI Systems, Jon serves as a Senior Consult for KMI/PAREXEL with specialties in computer system and cleaning system validation. Prior to joining KMI, Jon served as the Senior Manager of Validation at Biopure Corporation—a blood substitute drug manufacturer based in Cambridge, Massachusetts. Prior to working at Biopure, Jon was the Manager of Validation at Amgen, Inc., located in Thousand Oaks, California, where his responsibilities included validation of all manufacturing operations. Jon previously cofounded the Boston Area Chapter of the ISPE, and currently serves as the Chairperson for the PDA

Biotechnology Cleaning Validation Committee and as a member of the PDA Software Validation Committee. Jon graduated from Boston University in 1989 with a Masters Degree from the College of Engineering in Biomedical Engineering. He received his Bachelor of Science in Physiology degree in 1983 from the University of California at Davis.

CARMEN M. WAGNER

Dr. Carmen M. Wagner is currently the Director of Quality for Wyeth-Lederle Vaccines and Pediatrics in Sanford, North Carolina, a manufacturer of vaccines. Dr. Wagner has held a variety of quality-related positions in the health industry and has worked in the area of biological manufacturing for more than 10 years. She started her professional career as an Assistant Medical Research Professor at Duke University Medical Center in Durham, North Carolina. From academia, Dr. Wagner moved to small, and then large, pharmaceutical companies, including her tenures at E.I. DuPont and Johnson & Johnson. Her work experience includes positions in research, research and development, manufacturing technical support, quality assurance, regulatory compliance and validation. For the last 10 years, Dr. Wagner has worked on creating and improving quality systems that can help companies become more efficient in the transfer of new products/processes from development to operations. Dr. Wagner is an active member of the PDA and the ISPE.

VINCENT L. WU

Vincent L. Wu is Vice President of Integrated Biosystems, Inc., in Venicia, Calif. He was formerly a Group Leader in the Technical Services Department at Genentech, Inc., where he was responsible for the design and implementation of aseptic processes for the production of biopharmaceuticals. He has been involved in the development of novel processes for biopharmaceutical manufacturing, large-scale bulk freezing and thawing, and the development of a blow/fill/seal process for a protein product. Mr. Wu is the author of several publications and has been a speaker in the field on these topics.

1

INTRODUCTION

Kenneth E. Avis

University of Tennessee

Carmen M. Wagner

Wyeth-Lederle Vaccines and Pediatrics

Vincent L. Wu

Integrated Biosystems, Inc.

As biotechnology has moved from research and development to the production and marketing of biopharmaceutical products, greater emphasis on quality assurance (QA) and validation of the processes used has occurred. In fact, many new or improved technologies have been developed. Therefore, the inspiration for this book stemmed from a desire to help meet the need of biopharmaceutical professionals to keep abreast of and have access to a record of some of the developments that have occurred. Incorporating such developments and establishing their reliability are keys to obtaining speedy approval of new products and achieving compliance with current Good Manufacturing Practices (cGMPs) for the processes involved.

Foundational to research and the subsequent development of biopharmaceutical products is the purification, identification, and maintenance of bacterial and mammalian cell lines. At least as critical as proving the identity and purity of chemical molecules for the synthetic development of pharmaceuticals is establishing the reliability of the characteristics of the cells and cell lines for the growth of reproducible cultures in the production of biopharmaceuticals. Since the latter are living, the issues in achieving and maintaining purity and viability are even more challenging. Then, once the quality is known and assured, the validation of the processes developed must ensure reproducibility of the biological process and its outcome.

In most biotechnological systems, the objective is a specific protein or group of proteins. From a QA perspective, the protein(s) must be identifiable, of high purity, and reproducible. Potentially contaminating entities, such as viruses, endotoxins, mycoplasmas, and denatured product must be eliminated or inactivated. Concurrently, analytical methods must be developed to identify and quantitate the active constituent(s) and identify the absence of contaminants. Recent biological and instrumental developments have greatly enhanced analytical capabilities and the ability to validate both the production processes and the test methods. These and many other technological issues associated with the assurance of quality and the validation of the processes required to produce high-quality biopharmaceutical products are presented in this book.

Also of concern is meeting regulatory requirements in a changing technology. Before pharmaceutical and biopharmaceutical products can be made available to the public, they must meet the requirements of the U.S. Food and Drug Administration (FDA), as legislated by the U.S. Congress. Existing FDA requirements for the pharmaceutical industry are not entirely applicable to the biopharmaceutical industry, but those in place are legally enforceable. Efforts to modify the regulations to better meet the needs of both the pharmaceutical and the biopharmaceutical industries have been underway for some time. In 1997, the U.S. Congress passed the Food and Drug Modernization Act, the first significant change in the Federal Food, Drug, and Cosmetic (FD&C) Act in 35 years. Some of the changes specifically affect the biotechnology industry, a result due in part to the efforts of the Biotechnology Industry Organization (BIO). The amended act includes progressive changes, such as implementing a completely electronic submissions program after 5 years, eliminating the Establishment License Application, harmonizing procedures for the FDA Drugs and Biologics Centers, and a move toward eliminating the Center for Biologics Evaluation and Research (CBER) batch certification and monograph requirements for "well-characterized" biotechnology products.

Notably, Statute 830 will expand the FDA's current program to expedite the filing and approval of new therapies for serious or life-threatening conditions. It will also codify FDA regulations and practices designed to ensure that patients will have access to therapies for serious and life-threatening conditions before they are approved for marketing. This means that an increased number of clinical products for life-threatening conditions will potentially be made available to a larger number of patients sooner.

These sweeping changes are a result of both industry initiative and a recognition of increased confidence in the evolving industry— a strong endorsement of its scientific technology. With the roadblocks to drug approval decreasing, and with the streamlining of the approval process for drug manufacturing changes, there will be increased pressure for manufacturers to keep current with quality and compliance issues for both clinical and marketed products, including increased responsibility to accelerate timely collaborations with regulatory agencies.

As the industry and regulatory bodies worked together to define a well-characterized biopharmaceutical product, so should the effort continue to provide well-characterized manufacturing and QA procedures. The ability of manufacturers to effectively control and analyze the manufacture of biopharmaceutical products is dependent on the effective use of process control and its instrumentation, the development of specific assays for product and process contaminants, and the ability to demonstrate process repeatability through process validation.

This book presents a series of selected topics that define some of the unique challenges facing biotechnology companies in producing biopharmaceutical products. The topics selected address some of the quality and validation issues, starting with the cryopreservation of cell lines through the filling and finishing of the product. Further issues will be addressed in subsequent volumes. We believe you will find this book to be helpful to you in carrying out your responsibilities in biopharmaceutical processing.

CHAPTER CONTENTS

The following summarizes the somewhat divergent but related topics of the next six chapters. These topics were selected because of their timely nature, the current lack of coverage in available reference books, and the availability of highly qualified professionals to write on the topics selected.

Chapter 2, "Cryopreservation: Storage and Documentation Systems," is written by Frank P. Simione, who is Director of Professional Services and Safety Officer at the American Type Culture Collection (ATCC). Mammalian and microbiological cells are maintained in either private or public collections. The ATCC was established as such a public resource. The importance of the availability of viable, pure, and reliably identified cultures of cells needed for research and production in biotechnology cannot be overemphasized. The author has described how almost all cell lines are preserved for long-term storage by low-temperature freezing (cryopreservation) or by freeze-drying (lyophilization). Mr. Simione takes the reader through the various steps of cell preparation–their identification, preservation, and maintenance of the collection. The inventory and its control is essential in assuring the usefulness of the cells and their availability.

Chapter 3, "Quality Control and Quality Assurance Issues in Biopharmaceutical Processing," is by Dr. Gary L. Christiansen, a well-known consultant in the field. After reviewing the classical components of pharmaceutical quality control (QC) and QA, the author presents an extensive section on the specific concerns for biotechnology products. He discusses various test methods and requirements, including protein analysis; protein sequencing; peptide mapping; immunoassays; electrophoresis; release testing; and testing for safety, sterility, pyrogen, DNA (deoxyribonucleic acid), mycoplasma, and viruses.

The author then develops process validation concepts and requirements as related to biopharmaceutical processing, the importance of cell line characterization; the need for virus removal or inactivation; the removal of contaminating DNA and nucleic acids; and the removal of other contaminants, such as serum albumin, insulin, and denatured product. The last section of the chapter is an extensive coverage of the expectations of the FDA in an inspection, as identified in several published FDA guidelines. This chapter is an excellent and detailed treatise on QC and QA relative to biopharmaceutical products and their processing. The references include 28 citations, several of which are federal and international documents on regulatory issues.

Chapter 4 is written by Gregory Bobrowicz, an FDA investigator at the time of writing. The author begins "Biotechnology Manufacturing Issues: A Field Investigator's Perspective" with a brief description of the FDA's organizational structure, noting that of the five centers adjacent to Washington, D.C. (Rockville, Md.), CBER has primary responsibility for product approval for most biopharmaceuticals. A major portion of the chapter then reviews the topic of FDA concerns, related primarily to the safety of products for human administration. He discusses the occurrence of impurities and how their presence should be controlled, including viruses, mycoplasmas, microorganisms, endotoxins, and the general category of process contaminants. The author also discusses the FDA's concerns for potency and identity of the product.

In the next section, Mr. Bobrowicz provides a useful discussion of strategies for compliance with FDA requirements. He notes that quite often citations on 483s reflect a complete absence of an attempt at compliance. He also states that firms having good compliance records usually invest heavily in planning, data review, internal audits, managerial involvement, a corporate culture promoting quality, and, finally, an emphasis on investigating all deviations from prescribed limits. The chapter concludes with a comment on why the FDA expects manufacturers to practice adequate controls. This interesting and enlightening chapter concludes with a list of 63 references.

In "Validation of Biopharmaceutical Processes," written by Drs. Howard L. Levine of BioProcess Technology Consultants and Francisco J. Castillo of Berlex Biosciences, the authors examine the difficult task of validating a biopharmaceutical process. They emphasize the necessity of being sure that the right protein is obtained and maintained throughout the entire process, without introducing contaminating substances such as viruses, endotoxins, foreign proteins, media constituents, and process chemicals. They underscore the importance of validating the process so that it can be controlled and reliably repeated.

The authors have organized Chapter 5 by first discussing cell banks, the testing of specimens to be assured of freedom from contaminants and validation for cell integrity. They then discuss the operational qualification (OQ) of equipment, such as fermentors, incubators, media preparation equipment, sterilizers, biosafety hoods, chromatography systems, and filtration equipment. The fermentation or cell culture process is a critical one that must be run aseptically in a defined and controlled manner and must be validated repeatedly to produce the anticipated harvest of the product. The authors move from this topic to that of validation of downstream processing—the process of product purification. Downstream processing usually involves chromatography to purify the product and tangential flow filtration to concentrate the macromolecules.

In a discussion of the performance qualification (PQ) of downstream processes, the authors describe rigorous testing procedures to demonstrate the reproducibility of the process. Since biological processes typically introduce or fail to remove various contaminants, they must be subsequently inactivated or eliminated. Therefore, under the heading of clearance studies, the authors discuss the purification processes utilized and the testing required to ensure that purification has appropriately eliminated or removed such contaminants as DNA, host cell proteins, pyrogens, and viruses. The authors conclude the chapter with a discussion of tangential flow filtration and its validation—a very significant process in the purification of biopharmaceutical products. At the end of the chapter, the authors provide a very valuable list of 93 literature references from the field.

Chapter 6 is an in-depth examination of the topic of filtration to remove viruses from biological fluids. It was written by Dr. Hazel Aranha-Creado, a senior staff scientist at Pall Corporation. The first five sections of the chapter present a brief but classical review of background information concerning viruses—their sources and detection—along with methods of viral clearance and the regulatory approach to viral clearance. Since the issue of viral contamination is critical with respect to the safety of biological products, and viruses are potential contaminants in most biological products of human and animal origin, the processes must be designed and validated to give a high level of assurance that any viruses present have been eliminated by the processing conditions.

The author then discusses methodologies for virus removal from liquids by filtration, emphasizing both direct flow and tangential flow filtration. The various types of polymers used for the filter membranes are also presented. Since full dependence is placed on filters to effect virus removal, their integrity throughout their use in the process must be assured. Therefore, the author provides details of the testing available for determining filter integrity. The chapter concludes with an impressive list of 102 literature references.

Chapter 7, the final chapter in this volume, is a survey of the concepts and processes for cleaning equipment used in biopharmaceutical processing and the validation of cleaning processes. Jon Voss, president of KMI Systems, and Robert O'Brien, manager of validation at Biopure Corporation, have joined together to prepare their very important and practical chapter. The chapter is divided into two sections: The first section presents the concepts and methods for cleaning equipment after being used for biological processes, including their special cleaning requirements, while the second section presents the validation of cleaning processes. Both manual and automated clean-in-place (CIP) processes are covered in a helpful and practical manner in this quick-reference chapter.

It is believed that this book will find a welcome niche in biopharmaceutical processing technology as a reference tool because of the useful information it contains and the quality of its presentations. There is much to be learned concerning the relatively new field of biopharmaceutical processing. This book fills a significant number of gaps in the knowledge base for this specialty.

2

CRYOPRESERVATION: STORAGE AND DOCUMENTATION SYSTEMS

Frank P. Simione

American Type Culture Collection

The long-term preservation of biological specimens and associated data assists in ensuring reproducibility and comparability in biomedical research. There are two major goals in preserving living cells and organisms: maintain the specimens viable and unchanged for a long period of time and avoid contamination. Serial subculturing carries with it the risk of genetic change and exposure to unwanted contaminants. The cell culture literature contains numerous references to work performed with cells that were unknowingly contaminated, the most classic instance being that of HeLa contamination (Lavappa 1978; Nelson-Rees et al. 1981). This problem continues, and more and more sophisticated means of characterization have led to the discovery of other mixed or misidentified cell populations.

Low-temperature preservation by freezing and storage at cryogenic temperatures is the most effective means of maintaining cells and organisms unchanged for later use. Although the principles of low-temperature preservation are widely applied, the practices differ depending on the intended use of the stabilized cells. Effective preservation regimens start with the characterization and identification of the material to be preserved, since low-temperature preservation of cells is not a panacea, and one cannot expect to recover from the process better quality specimens than were present prior to preservation (Stevenson 1963).

The parameters for cryopreserving a living cell or organism are determined by the desired end use of the preserved material. Research material that is neither thoroughly studied nor well characterized is often preserved by low-temperature methods to ensure that it remains similar to the starting material. In this case, complete characterization of the specimens when preserved is not of great concern since these are often new isolates. However, identifying the specimens and being able to recover them when needed are important aspects of the preservation regimen. Production processes and quality control (QC) require the starting material to be both consistent in character and performance and free from all extraneous contaminants. When banks of cells and organisms are used in these processes, the characterization regimen prior to preservation and the inventory and storage procedures for maintenance must be highly controlled.

SELECTION OF MATERIAL FOR CRYOPRESERVATION

It is not economically desirable to cryopreserve every specimen encountered, and some mechanism must be established for choosing which specimens to bank. The selection of the desired material depends on the intended use of the specimens, and varies depending on the type of collection (Table 2.1).

Table 2.1. Types of Collections

Public Collections	Specimens available to the scientific community
	Specimens held in perpetuity
	Private specimens held under contract
Private Collections	Specimens held for institutional use only
	Specimens held in safekeeping in a public collection
	Specimens banked in compliance with regulations
Research Collections	Specimens maintained for single purpose
	Individual investigator controls inventory

Public Collections

Presently, there are more than 400 public culture collection agencies that make their collection holdings available to scientists for research and other uses (World Federation for Culture Collections 1993). These service collections have a mission to bank items useful to the scientific community and to provide continuity in their offerings so that scientists can compare their results with those of other investigators. This necessitates a focus on the maintenance of the entire genetic makeup of the specimen and ensuring the proper conditions for phenotypic expression when specimens are recovered, which requires the use of accepted preservation and inventory practices (Simione 1992b).

Acceptance of items into a public collection is contingent on their usefulness to the scientific community. The material is most often contributed by scientists who must demonstrate that the culture is useful to their colleagues. One means of accomplishing this is by publication in a peer-reviewed scientific journal. The culture collection will develop acceptance criteria, such as demonstrated usefulness, uniqueness to the collection, and capability for cryopreservation, that are then applied to each new item.

After accepting a culture into a collection, curators subject the new accession to a regimen of characterization tests to ensure that it is what the depositor claims it to be and that it is not contaminated. While many of these tests continue to be the traditional morphological and biochemical tests, more sophisticated molecular procedures have permitted greater scrutiny of new materials prior to promoting them for use in scientific research.

The preservation of cultures in a public collection requires a focus on maintaining the entire set of genotypic and phenotypic characters. Where a culture is useful for a specific purpose, such as the production of a specific antibiotic or antibody isotype, this characteristic becomes the focus of a good preservation regimen. To ensure that the cells remain unchanged for many years, public collections retain a subset of the initial lot of cryopreserved material for later use in replenishing distribution inventory (Simione 1992b).

Private Collections

Private collections are those containing living cells and organisms that are not available to the scientific community at large. These are maintained for the private use of an individual or institution and remain under the direct control of the owner. Private collections may be maintained by the owner of the material, or arrangements can often be made for a public culture collection to provide a proprietary caretaking service. Public collections often provide a mechanism for the maintenance of proprietary materials both for continued private use and in compliance with international regulations on depositing of materials for patent purposes.

The selection of material for private collections is contingent on the potential usefulness of the material for research or commercialization. When a cell or organism is a candidate for potential future use, the owner generally banks the specimen in a manner consistent with its intended use. The material may be placed in secure safekeeping at the owner's facility or entrusted to a public service collection that can provide proprietary, confidential, and secure maintenance. For cultures cited in a patent application, national and international regulations require that the material be deposited in a recognized patent depository under conditions that allow the depositor to restrict access to the material until the patent is issued, after which the depository must make the culture available to the scientific community (CFR 1997; WIPO 1994).

Cell banks developed for use in commercial processes are collections of specimens that must undergo stringent characterization and controlled preservation as part of an overall quality assurance (QA) program. Preserved specimens to be used in processes such as pharmaceutical or biological production must be banked in compliance with strict guidelines outlined by the U.S. Food and Drug Administration (FDA 1993). These private collections are only developed for highly valuable material and often require the owner to solicit outside assistance in producing, characterizing, and maintaining the bank.

Research Collections

Collections of material held for research purposes are generally not well characterized and are preserved solely as a resource for continued research. Research scientists are not always as careful about developing accurate and convenient inventories for their material. This often results in disorganized freezers, inadequate inventory databases, frustration when attempting to retrieve material from a freezer, and freezers of poorly identified material left behind when a research scientist retires or changes the focus of his or her research.

Since the preservation of cells and organisms is only a tool for ensuring the availability of these research materials, research scientists often pay little attention to the inventory and handling of these resources. Some institutions provide a centralized, controlled access facility for maintaining collections of research materials. While often frustrated when attempting to retrieve material from a centrally controlled facility, research scientists eventually find that their collection holdings are much better cared for; the scientists are relieved from the burden of monitoring the maintenance and inventory of their material. Central control of inventory can be provided by the scientist's institution, or material can be entrusted to a public collection.

Organizations with multiple investigators, each of whom manages his or her own inventory, are often faced with the problem of a widely dispersed and confusing inventory. This situation leads to redundancy, underutilization, and retention of unwanted material when an investigator leaves the organization. The conversion of a dispersed inventory into a centrally managed repository must be well planned. Criteria such as ownership of the material, access to specimens, and proper care of all specimens entrusted to the inventory must be carefully considered. Inventory systems must be designed to allow central repository staff to respond quickly and effectively to all requests for specimens.

The advantages of centralizing the control of frozen stocks of biological specimens include standardization of the inventory system, which allows wider usage of the stored material; controlled handling of the material by well-trained staff, eliminating the handling of frozen stocks by laboratory staff whose primary responsibility is not management of the inventory; and centralized control of security and safety of the inventory, including management of required resources and monitoring and responding to alarm conditions (Table 2.2).

CHARACTERIZATION

Identification and characterization of living cells and organisms are key components of a good cryopreservation and storage program (Table 2.3). It is difficult, if not impossible, to ensure that material remains unchanged during cryogenic storage if the key characteristics of the cells are not determined prior to preservation. A well-developed characterization program includes prepreservation analysis and sampling and key character assessment during storage to ensure stability and adequacy for postpreservation use. There is always the risk of change as cultures are handled and exposed to the rigors of the preservation process. Well-characterized material is a must in a good cryopreservation regimen, as it ensures through each step of the process that control of the effects of handling and

Table 2.2. Advantages of a Centrally Controlled Inventory

Research scientists are relieved of the burden of maintaining specimens in freezers.

Control of the inventory retained when investigators change research focus, relocate to another institution, or retire.

Lower costs due to consolidated support services such as maintenance, backup power, air-conditioning, liquid nitrogen supply, and inventory personnel.

Standardization of inventory and documentation systems.

Better security with centralized monitoring of alarm conditions, facility surveillance, and limited access to freezers.

Trained inventory staff to ensure proper handling of cryopreserved material using good safety practices.

Efficient handling of requests for materials, and better response in providing materials.

Table 2.3. Key Characteristics of Cells and Organisms

Identity:	Scientific name
	Source of tissue or cells
	Morphological characteristics
	Biochemical characteristics
	Isoenzyme and cytogenetic analysis
	Molecular analysis
Genetic Stability:	Wild type characteristics retained
	Mutants have not reverted
	Plasmids retained
	DNA not damaged
	Gene function retained
Purity:	Absence of other cells or organisms
	Absence of microorganisms
	Mycoplasma free
	Absence of extraneous viruses

preservation is maintained. The choice of characterization protocols depends on the nature and use of the material to be preserved.

Identity

Unidentified or misidentified specimens should never be cryopreserved. Numerous methods are available for identifying cultures; for most microorganisms, standard protocols can be used to characterize the material and assign it to a genus and species. Confirmation of identification can be accomplished by using molecular techniques such as DNA (deoxyribonucleic acid) fingerprinting or specialized methods such as gas chromatography, fatty acid analysis, or cell wall analysis. There are, however, instances in support of research protocols where unidentified specimens may be preserved and maintained for future use. These specimens should always be considered candidates for destruction at the completion of the study, as they will most likely never be useful outside the specific research protocol.

The authentication of mammalian cells is critical since cells from different species or tissues can often look alike. Confirmation of species, or tissue specificity, is accomplished by isoenzyme analysis, cytogenetic analysis, or DNA fingerprinting. The misidentification of cells can lead to wasted research efforts.

Viability and Utility

Two key parameters for the successful cryopreservation of living cells and organisms are retention of viability—the ability to reproduce—and the key qualities of the cells. In order to begin a biological process involving a cell or microorganism, viable starting material must be available for scale-up to production quantities. The key process quality of the material must be present, or the process will not take place.

Most efforts in cryopreservation are directed toward the recovery of viable cells or organisms. Testing for viability following freezing and low-temperature storage can be performed by several means, including cultivation in the growth medium of choice or by vital staining. Cultivation, with growth and reproduction, is the only method of certifying the recovery of viable cells following preservation.

Contamination

The contamination of cultures of living cells and organisms can be a devastating experience, especially if the contamination is subtle, such as mycoplasma contamination of cell cultures. The more frequently and extensively cultured cells and organisms are handled, the greater the chance for contamination. For some sensitive cells, the freezing process can actually result in selective separation of the contaminant, which is often more resistant than the desired material.

Tests for contamination depend on the type of contaminant and the nature of the specimen being preserved. For example, microorganisms can be cultured on solid media, where the separation of colonies allows a visual determination of purity. More subtle contamination of cultures by microorganisms is detected by biochemical or molecular characterization methods. Cultures of mammalian cells are subject to contamination with bacteria, fungi, viruses, and mycoplasma. Bacterial and fungal contamination will generally result in visual changes in the culture. However, contamination with viruses or mycoplasma is not obvious, and special testing is required to detect these cryptic agents (Hay et al. 1992).

Only some mycoplasma species can be cultured for direct detection. Therefore, several methods are used for the detection of mycoplasma, including direct detection by cultivation on agar and in broth and indirect detection by bisbenzamide DNA fluorochrome staining. Molecular techniques have also been developed for the detection of mycoplasma using the DNA polymerase chain reaction (PCR) (Rawadi and Dussurget 1995).

Genetic Stability

Maintaining the genetic stability of cells and organisms is a key challenge in the preservation process. All living organisms tend toward change, and mutations and reversion to the wild type can occur. To ensure that change does not occur, the testing of cells and organisms following preservation is necessary. Sophisticated methods are available for assuring the genetic stability of cells and organisms (Hay et al. 1992). Cytogenetic analysis by chromosome banding and marker identification is used to ensure that populations of mammalian cells are uniform and remain unchanged. Isoenzyme analysis is used to identify cells to species and can be used to identify the tissue type as well. DNA fingerprinting and other molecular techniques are now routinely used to ensure the stability of cells and organisms.

Cell Banking

The production of biological products subject to licensing with the FDA requires that the cells or organisms used in the process be well characterized and maintained under stringent control (FDA 1993).

One of the major concerns is that components of the cell system used may be carried into the final product. Therefore, safety testing must be performed on the system, which includes testing for viral, nucleic acid protein, and even trace compound contaminants. Stabilization of the cell system minimizes variability in the final product due to changes in the cell system.

Development of master cell banks (MCBs) and manufacturers working cell banks (MWCBs) ensures consistency in the final product. The MCB is derived from a single tissue or cell line; preparation of the bank requires strict adherence to current Good Manufacturing Practices (cGMPs). All aspects of the development of a cell bank must be well controlled and documented. Cell banks are stored in liquid nitrogen freezers under carefully controlled conditions that include monitoring of the freezer temperature, accurate inventory control, and redundant backup by separating the cell bank into two freezers and storing a portion of the bank off-site.

Preservation

The freezing of cells and organisms is a complex process that has been studied enough to allow the use of standardized protocols. Phenomena during the freezing process include ice crystal formation and subsequent changes in solute concentration inside and outside the cell. The addition of cryoprotectants and lowering of the temperature at controlled rates have resulted in established protocols for the successful preservation and recovery of frozen cells. After freezing, cells must be maintained at a temperature that prevents conformational changes due to recrystallization of ice (Simione 1992b).

A good cryopreservation protocol includes an assessment of the optimum growth conditions of the material to be frozen. The better the condition of the cells at the beginning of the process, the more resistant they will be to the rigors of freezing. The growth medium, incubation conditions, and the state of the life cycle of the cells or organisms all contribute to successful recovery following freezing.

The addition of a cryoprotectant such as glycerol or dimethylsulfoxide (DMSO) is necessary in most situations to ensure adequate protection of the cells during freezing (Simione 1992a). Cryoprotectants primarily serve to buffer changes in solute concentration during the freezing process; in some cases, they may even protect against physical damage from ice formation. The chemicals used as cryoprotectants can sometimes be toxic to the cells. Therefore, the concentration of the cryoprotectant is critical, and exposure of the cells to the cryoprotectant prior to freezing should be minimized. Glycerol and DMSO, in concentrations of 5–10 percent v/v, are commonly used to preserve most cells and organisms, and a period of incubation with these additives prior to beginning the cooling process is important to ensure adequate exposure of the cells for maximal protection. However, the exposure time should be no longer than 30 min prior to beginning the freezing process.

Freezing of the cells is accomplished by placing the specimens in vials or ampoules in 0.5 to 1.0 mL portions and slowly cooling the cells at a rate of approximately 1.0°C/min to approximately –40°C. Many cells and organisms will tolerate an uncontrolled cooling regimen, such that they can be placed in a freezer at –70°C, where the ambient temperature results in more rapid and less-controlled cooling of the specimens. For more sensitive cells, a controlled rate of cooling is required that also includes a means of minimizing the latent heat of fusion during ice formation (Simione 1992b).

Following controlled cooling, the specimens must be immediately placed in a freezer at the proper storage temperature. The critical temperature for guaranteeing long-term stability of sensitive living cells and organisms is below –130°C (Simione 1992b), because changes in the conformational properties of ice can continue to occur above this temperature. To ensure the frozen specimens are maintained below –130°C, the freezers must be designed to operate below this critical temperature at all times during storage and retrieval activities and in all locations within the freezer (Simione and Karpinsky 1996). In practice, freezers should be set up to maintain a temperature of –150°C or below.

Mechanical freezers do not provide the stable environment required for most sensitive cells during frozen storage, although in some cases they are designed to reach temperatures of –130°C to –150°C (Mikoliczeak et al. 1987). Liquid nitrogen freezers are the only reliable means of ensuring an adequate working temperature for sensitive cells; in every instance, the design of the freezer and the inventory configuration are critical for ensuring a constant temperature.

Freeze-drying of microorganisms can be an alternative to freezing, especially when material must be used in situations where cryogenic storage is not practical or when material must be shipped. Culture collections that regularly ship items as part of their service utilize freezing-drying as a means of preparing specimens that are more economical to maintain and distribute (Simione 1990). However, the freeze-drying process cannot be used for all specimens, only for bacteria, some fungi, and some viruses. The process is expensive, and the material generally is not as stable as cryogenically stored specimens.

INVENTORY

The type of inventory system chosen depends on the specimens to be maintained and the intended use of the preserved material. The physical configuration of the inventory and management of the supporting data require thorough preplanning and assessment before designing the inventory and data management systems. This is probably the most critical step in setting up a bank of specimens for research, production, or public use.

In public collections, specimens are assigned unique identifiers, or accession numbers, that are used to track the items through all subsequent processing. When these items are stored, locator information is based on the unique number and a batch number for each production lot. Since all material in a production batch is the same, when a batch is removed from the inventory, the entire complement of vials making up the production batch is removed. This is also the practice in private collections held for proprietary and confidential use.

When specimens are collected for research or patient studies, each individual specimen is a unique inventory item. When a research study involving the collection and maintenance of specimens is developed, part of the overall plan must address the inventory of preserved specimens. Research specimens are often assigned unique identifiers, but they are seldom expandable into batch quantities of like items. Therefore, when a research study is nearing completion or enters a new phase, the need to remove unwanted items from the inventory could be a daunting task. Preplanning that allows some segregation into separate freezers, or batches of specimens within a freezer, will dramatically reduce the task of later removing unwanted specimens from the collection.

General guidelines for inventory management have been established and should be followed for all cryopreserved cells and organisms (Simione 1993):

- Divide the inventory into more than one portion, maintained in different locations, to provide for well-characterized starting material for new lots and backup material for safekeeping.

- Use a physical storage system that allows convenient access to all material to avoid unnecessary exposure of workers to extremely cold temperatures and unwanted exposure of frozen inventory to warmer temperatures.

- Control access to the inventory and, if practical, require multiple investigators to maintain their material in a central location under the control of one individual or unit.

- Use lot identification and locator systems that are easy to understand and require all investigators to comply with the standard systems.

- Ensure that adequate and dedicated electrical power is available and that all freezers are backed up by emergency electrical power. If economically practical, spare freezers should be available for relocation of inventory in the event of primary freezer failure.

- Install an alarm system that allows constant surveillance of freezer conditions and a warning system that is designed to alert staff anytime an alarm condition occurs.

- Install a temperature monitoring system that allows constant monitoring of the temperature in the warmest section of the freezer.

- Ensure that all personnel working in the freezers are trained to understand the importance of maintaining the critical temperature of all items at all times until permanently removed from the inventory.

Segregated Inventory

The importance of a segregated inventory for ensuring secure storage and retrieval of stabilized material cannot be overemphasized. There are two important components to a segregated inventory:

1. Segregation to ensure that starting material for new lots has not changed.

2. Segregation to protect from physical loss of all material in a lot.

Segregated starting material is called seed stock in culture collections and consists of low-passage material preserved from original specimens. Proper maintenance of seed material ensures that future production lots of the specimen are unchanged when compared to the original starting material. Segregation to protect from accidental loss due to freezer failure or other physical disasters consists of maintaining duplicate items in a separate physical location. These specimens can be maintained in separate freezers at the same location or in a separate facility, such as a public collection that offers a safekeeping service.

The goal in stabilizing living material by low-temperature preservation is to ensure that changes in the specimens do not occur during storage. The time and effort invested in establishing a low-temperature storage system is wasted if the ultimate goal of maintaining the material unchanged is not achieved. The most elaborately designed inventory system will not protect cells and organisms from change if proper procedures for handling are not established, and inventory staff are not adequately trained. The key element in establishing handling procedures is to ensure that specimens are maintained below the critical storage temperature at all times until retrieved for final recovery.

Since subculturing is required to produce material for new frozen lots, each new lot may differ slightly from the previous one. Continuation of subculturing can lead to shifts in the characteristics of the stored material unless controlled production and adequate QC procedures are established. To protect against changes that may occur during the production of new lots, a seed lot should be established that can be used as starting material for new working lots (Simione 1992a, 1992b). To do this, the initial lot of a well-characterized cell or organism should be divided into seed and working stock. The seed stock is held in reserve for use only to replenish the working stock. When additional working stock is needed, a sample of the seed material is retrieved for use as starting material for the new lot. In culture collections that have maintained material for many years, the seed stock system has allowed maintenance of material that is only a few passages from the original starting material. Seed stock should be maintained in a separate freezer from the working stock.

Even a well-designed inventory system is subject to potential failures. The best means of minimizing the risk of loss of material in any system is redundant backup. A portion of each lot, preferably seed stock, should be set aside and secured for safekeeping at a physical location separate from the main inventory. While maintaining the segregated material in a separate freezer in the same facility provides some assurance, an off-site location is the best choice. The off-site location must provide the same safeguards as the primary location, such as backup electrical power and alarms. Many public culture collections provide a service for proprietary and confidential maintenance of valuable cultures in their facility for safekeeping.

Physical Inventory

Inventory systems should be chosen that are compatible with the primary storage vessel. For vials and ampoules, a vertical stacking system is the least risky, as a specimen at the top can be retrieved without exposing any material below it. If a horizontally arrayed inventory is desired, such as in a box storage system, the boxes should be stacked in a manner that ensures material intended for long-term storage is located at the bottom.

When retrieving specimens from cryogenic storage, only those items desired for immediate use or transport should be removed from the freezer. Never return material to a freezer once it is removed, even for short periods of time. Samples exposed to warmer conditions can experience a rapid rise in temperature. Repeated exposure to temperatures above -130°C can result in the loss of viability of sensitive material.

The extreme temperature conditions experienced while working in liquid nitrogen can lead to handling practices that compromise specimen temperature and expose personnel to the hazards associated with immersion of material directly into liquid nitrogen. When specimens are immersed directly in liquid nitrogen, personnel must be adequately protected from the cryogenic liquid to avoid skin exposure or splashing on the face. This requires adequate personnel protection, including a face shield and insulated gloves. Wearing well-insulated gloves compromises the ability to handle small specimen containers, such as cryovials, and may lead to undesired exposure of the specimen to warmer temperatures or the dropping of specimens.

The less extreme environment in the vapor phase of a liquid nitrogen freezer lessens the problems associated with personnel protective devices, but greater attention to proper handling practices is even more critical. Handling specimens in the vapor phase of a liquid nitrogen freezer can result in a greater risk of specimen exposure to warmer temperatures unless the freezer is properly validated, and handling procedures are well developed and strictly adhered to (Simione and Karpinsky 1996).

Physical inventory systems for cryogenic storage should be designed to minimize the handling of material. While some systems are more efficient than others, there are principles that can be applied to any cryogenic storage system. Since it is not practical to store each lot of frozen stock in a separate freezer, the inventory system should be structured to allow proper segregation of specimens and easy access to any component of the inventory without exposing any other portion to warmer temperatures. Inventory control systems in mechanical freezers are usually designed to allow the storage of small boxes of materials in vertically arrayed racking systems. Individual boxes can be used for similar material, allowing access to a particular component of the inventory via retrieval of an individual box. The advantage of a box storage system is that any size container that will fit inside the box can be used as the primary container for the frozen stocks. The principal disadvantage of this system is the necessary exposure of material stored at the top of the racking system while attempting to retrieve material at the bottom.

If a box storage system is used, some organized means of storage should be designed to ensure that frequently retrieved material is maintained near the top of the freezer, such that the longer an item remains in the freezer, the less it is exposed to warmer temperatures. Therefore, it may be desirable to separate a batch of material into two portions, maintaining some of the material at the top of the racking system, and the remainder at the bottom. For frequently accessed sensitive material, this will ensure that at least a portion of the inventory is below the critical temperature at all times, since the operator will consistently retrieve material only from the top boxes. If this constant handling of the top boxes results in loss of titer due to temperature fluctuations, unchanged specimens can be retrieved from the duplicate noncompromised inventory in the bottom boxes.

While mechanical freezers are useful for frequently used material, sensitive items should not be maintained in these freezers. Generally, frozen stocks of cell cultures and hybridomas, as well as sensitive microorganisms, should always be maintained in a liquid nitrogen freezer. However, using a box racking system in a liquid nitrogen freezer presents the same exposure concerns as for a mechanical freezer, and the same guidelines should be followed in managing the inventory. Therefore, the system should be designed to allow operators to retrieve material from the top boxes only, preserving duplicate material in the bottom boxes as backup. As discussed below, liquid nitrogen freezers have a more stable temperature gradient, which lessens the potential for exposure to warmer temperatures during stocking and retrieval operations.

The ideal system for the storage of living materials is in liquid nitrogen freezers designed for use with a cane storage system (Simione 1992a, 1992b). This inventory system consists of clipping six vials containing the same material in a vertical array on aluminum storage canes (Figure 2.1). The canes are labeled with the specific location in the freezer, and the vertical array allows the operator to remove only the top vial without lifting the remaining vials above the critical temperature space. For the storage of frozen cells and organisms, this system only works if the primary storage container is a standard diameter that can be accommodated by the cane storage system. During the development of a cryogenic storage system, consideration should be given to utilizing standard cryovials of

Figure 2.1. Inventory system using aluminum storage canes for vials.

1.0 to 2.0 mL capacity for all frozen specimens and storing them in a cane system.

The cane system can be adapted to any liquid nitrogen freezer; however, the newer and more efficient units designed to maximize capacity and temperature stability provide the most useful configuration for cane storage.

There are three major advantages of the cane storage system (Table 2.4):

1. More of the available inventory space in the freezer is occupied by the primary specimen containers, in contrast to the

Table 2.4. Cane Versus Box Inventory System

Canes	Boxes
Vials in vertical array.	Vials in horizontal array.
Remove one vial at a time.	Remove entire box.
Lift only one, six-vial, cane.	Lift entire rack of boxes.
Only retrieved vial is exposed to warmer temperatures.	Entire array of vials is exposed.
Only one vial is handled.	Entire box of vials is handled.
Minimal personnel exposure to cold temperatures.	Sustained personnel exposure to cold temperatures.
Risk limited to dropping of one vial.	Risk includes potential for dropping numerous vials.
Ratio of inventory system occupied area to vial area is small (larger vial capacity).	Ratio of inventory system area to vial area is larger.
Cane location can be reused after only six vials are removed.	Box location houses up to 100 vials.

    box system where empty box space, dividers, and walls utilize some of the available space.

2. By storing the primary containers in a vertical array, the top specimen can be retrieved without exposing the material below it to warmer temperatures. It is important that identical specimens be maintained on a single cane, and that the cane not be raised any higher than necessary to grasp the top specimen.

3. When each cane is labeled with a specific identifier, the need to search for a specific item among many is eliminated. This minimizes the exposure of unwanted material to warmer temperatures and the exposure of operators to the extreme cold temperatures encountered in a cryogenic storage system.

Inventory Access

Access to the inventory is often overlooked when setting up an inventory control system. Allowing uncontrolled access results in lost items, mishandling of material, and inaccurate inventory records.

Institutions should consider centralized control of their inventory of frozen stocks and should place that control in the hands of a few qualified individuals. This ensures standardized inventory control and handling and provides users with the maximum security for their valuable materials.

The centralized inventory should begin with a single location within the facility with limited access. Access can be controlled by lock and key or card access and should be limited to those individuals authorized to enter. All authorized staff should be adequately trained in the utilization of the inventory system and in the proper handling of cryogenically stored living materials. Owners of the material must rely on the authorized inventory staff to handle their specimens and should work with the inventory staff to ensure proper control and handling.

Access to individual freezers may need to be controlled as well. This can be accomplished by locking the freezers and controlling access to the locator codes for the freezers. In a facility where general access is allowed for research materials or specimens certified for distribution, it is imperative that proprietary material not available for general access be secured. In culture collections where services such as maintenance of restricted patent deposits and storage of valuable material for safekeeping are offered, it is critical that these items are kept separate from the general distribution inventory. Research materials should be segregated from material used for production and QC testing.

A well-designed inventory system requires preplanning to guarantee standardized coding of specimen identification and location. Individual investigators and owners of material should not be allowed to use personal identification codes. Previously used personal identifiers can be cross-referenced to a standardized code, the latter permitting maintenance in a centralized inventory system. Each specimen must have its own unique identifier, and each lot of the inventory must be identifiable and traceable. The most convenient system for unique identification is a numbering system that is applied sequentially for like material, coupled with an identifier for the storage facility.

In large collections, such as the American Type Culture Collection (ATCC), items are assigned unique catalog numbers that are retained in perpetuity. This means that the identifying code is retained without change and is retired in the event the item is removed from the inventory. The facility identifier for ATCC cultures is the "ATCC" acronym. Each collection of material, such as bacteria, fungi, and hybridoma, is assigned a set of catalog numbers to be used for new items assigned to that collection. However, all ATCC items automatically carry the "ATCC" acronym as part of the catalog number, identifying that item as part of the ATCC collection.

Locator codes should be assigned in a manner that allows easy identification and retrieval. Each freezer should be assigned a unique number; depending on the type of freezer, sections within the freezer can be identified by number or letter code as well. When boxes are used, the system should include a unique number or letter code for each box within the freezer and the section of the freezer where appropriate.

Specimens stored in boxes in horizontal racking systems, which are commonly used in chest type mechanical freezers, can be assigned locator codes in the following manner. The first number or letter should identify the freezer, the second number or letter should identify the rack, and the last identifier should refer to the box. The inventory within an identified box can then be located by placing items into a grid with a conventional numbering system. For example, a freezer may be identified as number "5,", the rack as number "1" (often Roman numerals or letters are used), and the boxes identified as number "1" through "6" (or the corresponding number of boxes in the rack). The code then appears as 5:1:1, 5:I:1, or 5:A:1 in the inventory control system.

When a cane storage system is utilized, the following notations can be used. The freezer is again numbered (e.g., freezer "5"), the section of the freezer in which the cane is found (e.g., "1"), the holding box in which the cane is located (the convention is 16 canes per box; e.g., box "1"), and the individual cane (e.g., cane "1"). In this system, the code appears as 5:1:1:1 in the inventory control system.

These coding systems are simple enough that the entire locator code can be printed on the individual box or cane for ease of location during retrieval and stocking operations. These codes also allow restocking of different items in a location vacated by previous items and, when properly maintained, provide a means of identifying what is contained in any freezer location.

Security

To ensure the security of frozen stocks, steps must be taken to provide redundant support in the event of a failure of the operating system(s). Adequate electrical power must be available to all freezers at all times, and the entire system should be backed up by emergency electrical power. Testing of the emergency power system should take place weekly to ensure that backup power is available when failures occur.

An automated freezer monitoring system is necessary for all low-temperature storage facilities. Each freezer should be equipped with an individual alarm monitoring capability that senses temperature changes or liquid level fluctuations in liquid nitrogen freezers. Temperature-sensing monitors should be installed that operate separately from the freezer sensors. These sensors should be monitored by a facility-wide surveillance system that detects changes in temperatures and provides a facility-wide alarm. In large facilities, it is important to design a system that allows pinpoint identification of the problem at a specific freezer. Staff should be trained and assigned to respond to facility-wide alarm conditions, and the system should be monitored by an off-site central station at all times for redundant notification of staff when an alarm condition occurs.

The training of staff is discussed in greater detail below; however, it is important that a mechanism for response be established. Key, well-trained staff should be assigned response duty such that when an alarm occurs, it is clear who is to respond and what their role in the response should be. Facility engineers can be assigned to respond to alarm conditions, but they should be instructed to contact the inventory manager or other staff adequately trained to handle the frozen material in the event freezer contents must be moved. All staff on the response team should carry pagers and be accessible at all times when alarm conditions occur.

Finally, spare freezers, with adequate space to accommodate the inventory of the largest freezer in the facility, should be available when problems occur. An adequately designed inventory system, as described above, must allow tracking of the location of any item in the inventory in the event material is relocated.

DATA MANAGEMENT

Good data management is critical to any low-temperature inventory system and includes the ability to cross-reference characterization data with specimen identifiers, lot numbers, and locator codes. There are many methods available for managing the data, and small collections may choose a computer (PC)-based system. Larger collections should develop an integrated database that ensures data generated from several sources are properly codified and cross-referenced.

The centralized control of inventory data must allow for several means of accessing the data. The unique identifier–catalog number–is the key starting point for all data retrieval. The system should be designed to ensure that all data generated on an inventory item can be tracked using the unique identifier, including lot number. When the laboratory generates characterization data and enters it into the integrated database, the unique identifier for that material should be used to ensure the data are properly assigned in the system.

A good disaster recovery system will provide for redundant backup of all data in the event the primary source is lost or destroyed.

REMOVAL OF COLLECTION MATERIALS

A complete inventory management system should include a mechanism for reviewing collection holdings and ensuring that the inventory does not become unwieldy. Planned, periodic review of the usefulness of holdings is essential. This is especially important for those facilities that manage the inventory for a number of users. Research scientists who retire or change the focus of their research often do not review their holdings and remove material that is no longer needed. If new material is added, and the older material remains, the collection can soon become economically unstable.

Inventory managers are constantly faced with the problem of storing large quantities of items for which there appears to be little use. The inventory system must include a mechanism for monitoring usage; when activity diminishes for a set of stored items, the status of the material must be reviewed. Large public collections are faced with the requirement of maintaining archival materials that may be of use in the future and must design a system that allows retention of these items in a manner that does not compromise storage and retrieval of the popular items. Smaller research collections may choose to remove unused items that are no longer of importance.

HANDLING AND DISTRIBUTION
OF COLLECTION HOLDINGS

Most collections of frozen cells and organisms are maintained for some purpose. This means there is a constant need to retrieve samples and provide them to the end user. The retrieval and distribution of frozen stocks requires the establishment of procedures that ensure the integrity of the material during retrieval and transport, including the proper containment of hazardous materials when these are handled.

The retrieval of frozen stocks from low-temperature storage requires handling procedures that ensure that the retrieved item is not warmed until ready for use, and that other items remaining in the freezer are not exposed to warmer temperatures. All handling of frozen stocks should be performed inside the freezer and close to the area where the stocks are stored. This ensures that the handled items are not at risk of warming.

Under no circumstances should a box be removed from a freezer to be opened and searched. Once a box is identified, it should be placed at the top of the freezer near the opening when opened and searched. If searching does require removing a box from the freezer, it should be returned to the freezer as quickly as possible.

Once retrieved from frozen storage, specimens must be kept frozen by placing them in liquid nitrogen–containing transport containers or in dry ice for temporary transport. Packaging and labeling regulations must be observed, and there are specific requirements for shipments using dry ice or liquid nitrogen.

TRAINING

All personnel who handle specimens of cryogenically stored sensitive materials must thoroughly understand the risks to the stability of the material as a result of improper handling. Training in cryobiological principles, including the importance of the maximum critical storage temperature, is essential to an understanding of proper handling procedures. Emphasis must be placed on the importance of ensuring that samples remain at or below the critical temperature at all times until retrieval for warming and use.

The purpose of training is twofold: to understand the design and operation of freezers and to recognize the importance of design criteria to adequate freezer operation. Some personnel may be trained solely in use of the freezers; others may need training in operation and maintenance of the freezer. All users and operators must understand the basic operating principles and the critical functions of the units. Power conditions and alarm monitors must remain active at all times. Degassing, a means of clearing the working area of the freezer, is commonly performed by activating a solenoid valve, but staff must clearly understand the potential impact of too frequent or prolonged degassing.

Facilities in which liquid nitrogen is piped from a stand tank to the individual freezers may have one of two designs. In older systems, a means of bleeding the piping to remove trapped vapor and allow flow of liquid only to the freezers is necessary during filling operations. Therefore, when degassing a freezer during inventory handling activities, vapor only enters the freezer, resulting in boiling of the contained liquid. Too frequent degassing can result in a lowering of the liquid nitrogen level; in all-vapor freezers, this will compromise the temperature gradient. In newer systems, the piping is designed to provide liquid nitrogen to the freezers on demand; in this instance, too frequent degassing may result in overfilling. In all-vapor freezers, this could result in exposure of material at the bottom of the inventory to liquid nitrogen.

It is important to ensure, through adequate training and retraining, that inventory staff understand the potential effects on both personnel and sensitive specimens of prolonged periods of working in the freezers. The addition of a large volume of warm materials to a freezer may cause a disruption of the temperature gradient; in mechanical freezers, this often results in an alarm condition. Staff must also be cautioned about the susceptibility of vacuum-insulated freezers to physical damage, such as the potential for puncturing the thin internal lining critical to continued insulation. Freezer lids and inventory racking systems must be handled gently and must be replaced immediately when work in the freezer is complete.

Staff who are assigned to respond to alarm conditions must also be trained at a level commensurate with their expected action. Engineering staff must be aware of the need for immediate response and the hazards associated with cryogenic gases. They should be trained to monitor and respond to alarm conditions, troubleshoot the problem, and respond (if necessary) with mechanical corrections. Those staff assigned overall responsibility for liquid nitrogen freezer maintenance must thoroughly understand the mechanical operations, alarm system operation, and controls and must be familiar with the freezer contents and how to handle them. However, engineering staff should not be responsible for relocating sensitive and potentially hazardous biologicals.

Training in proper safety procedures is essential for all personnel. This includes instruction in using proper protective clothing, such as gloves and face shields, as well as how to respond to emergency situations, such as low-oxygen alarm conditions.

PERSONNEL SAFETY

Several safety issues must be considered in the design and operation of a liquid nitrogen facility. Cryogenic gases present a hazard because of their extremely low temperatures; as warming occurs, gases such as nitrogen can displace oxygen in the work environment. Improperly ventilated containers of cryogenic gases, as well as improperly sealed specimen containers exposed to liquid nitrogen, can result in explosive hazards.

Personnel working in liquid nitrogen freezers must be provided with proper protective clothing, such as a laboratory coat, gloves, and face protection. Materials stored in liquid nitrogen freezers are extremely cold and can result in severe burns if handled without adequate thermal protection. When working in the liquid nitrogen itself, protection from splashing must be guaranteed, and adequate precautions must be taken when handling sealed vials that have been immersed in liquid. Operators must be protected from the potential of exploding vials. Use of all-vapor nitrogen storage removes the explosion potential associated with storage in liquid nitrogen; hazardous materials should never be stored directly in liquid nitrogen.

Nitrogen gas can quickly displace oxygen. Since it cannot be easily detected, precautions must be taken to warn personnel of diminished oxygen levels. Liquid nitrogen should be handled in well-ventilated areas, and facilities should be equipped with oxygen monitors with adequate warning systems; personal oxygen monitors for all staff should also be considered. Liquid nitrogen should only be placed in well-ventilated portable containers for transport or temporary use.

REFERENCES

CFR. 1997. *Biotechnology invention disclosures, patents, trademarks and copyrights.* Title 37, Part 1, Subpart G of the Code of Federal Regulations.

FDA. 1993. *Points to consider in the characterization of cell lines to produce biologicals.* Rockville, Md., USA: Food and Drug Administration, Center for Biologics Evaluation and Research.

Hay, R. J., J. Caputo, and M. L. Macy. 1992. *ATCC quality control methods for cell lines.* Rockville, Md., USA: American Type Culture Collection.

Lavappa, K. S. 1978. Survey of ATCC stocks of human cell lines for HeLa contamination *In Vitro* 14:469–475.

Mikoliczeak, J., M. Stephens, and F. P. Simione Jr. 1987. Comparison of –130°C mechanical refrigeration with liquid nitrogen storage for biological specimens *SIM News* 37:74.

Nelson-Rees, W. A., D. W. Daniels, and R. R. Flandermeyer. 1981 Cross-contamination of cells in culture. *Science* 212:446–452.

Rawadi, G., and O. Dussurget. 1995. Advances in PCR-based detection of mycoplasmas contaminating cell cultures. *PCR Methods and Applications* 4:199–208.

Simione, F. P. 1990. Freezing versus freeze-drying for the preservation and distribution of specimens from culture collections: A compact freeze-drying course on the fundamental aspects of the preservation of sensitive biologicals. Course cosponsored by the the Merieux Foundation and the American Type Culture Collection, 2–4 April, in Washington, D.C.

Simione, F. P. 1992a. *Cryopreservation manual.* Rochester, N.Y., USA: Nalge Company.

Simione, F. P. 1992b. Key issues relating to the genetic stability and preservation of cells and cell banks. *J. Parent. Science Technol.* 46:226–232.

Simione, F. P. 1993. Cell storage facilities and inventory. In *Cell & tissue culture: Laboratory procedures,* edited by A. Doyle, J. B. Griffiths, and D. G. Newell. New York: John Wiley & Sons, 1C:1.1–1C:1.4.

Simione, F. P., and J. Z. Karpinsky. 1966. Points to consider before validating a liquid nitrogen freezer: Validation practices for biotechnology products. STP 1260. Philadelphia: American Society for Tesing and Materials, pp. 24–30.

Stevenson, R. 1963. Collection, preservation, characterization and distribution of cell cultures. *Proceedings of the Symposium on the Characterization and Uses of Human Diploid Cell Strains.* Opatija, Yugoslavia: International Association of Microbiological Societies.

WIPO. 1994. A guide to the deposit of microorganisms under the Budapest Treaty. Geneva: World Intellectual Property Organization.

World Federation for Culture Collections. 1993. *World directory of collections of cultures of microorganisms,* 4th ed. Saitama, Japan: WFCC World Data Center of Microorganisms.

3

QUALITY CONTROL AND QUALITY ASSURANCE ISSUES IN BIOPHARMACEUTICAL PROCESSING

Gary D. Christiansen

Validation and Compliance Services

It is a requirement of the current Good Manufacturing Practice (cGMP) regulations that manufacturers of medical products, including both drugs and devices, have a quality assurance (QA) program. A variety of specifications, controls, and programs must be established and practiced to ensure that the products meet the established specifications. The cGMP regulations are based on fundamental QA concepts/principles. Quality, safety, and effectiveness must be designed and built into a product–it cannot be inspected or tested into a finished product. Each step of the manufacturing process must be controlled to maximize the probability that the finished product will be acceptable and meet its specifications. These requirements also apply to the biopharmaceutical industry; however, there are a number of issues for the biopharmaceutical industry that require special attention due to the nature of the products and the processes used to produce the products.

The specific regulations (Code of Federal Regulations 1995) that require a quality program for drug products are in 21 CFR 211.

211.22 Responsibilities of quality control unit

(a) There shall be a quality control unit that shall have the responsibility and authority to approve or reject all components, drug product containers, closures, in-process materials, packaging material, labeling, and drug products, and the authority to review production records to assure that no errors have occurred or, if errors have occurred, that they have been fully investigated. The quality control unit shall be responsible for approving or rejecting drug products manufactured, processed, packed, or held under contract by another company.

(b) Adequate laboratory facilities for the testing and approval (or rejection) of components, drug product containers, closures, packaging materials, in-process materials, and drug products shall be available to the quality control unit.

(c) The quality control unit shall have the responsibility for approving or rejecting all procedures or specifications impacting on the identity, strength, quality and purity of the drug product.

(d) The responsibilities and procedures applicable to the quality control unit shall be in writing; such written procedures shall be followed.

The specific regulations (Code of Federal Regulations 1995) that require a quality program for medical devices are in 21 CFR 820.

820.20 Organization

Each manufacturer shall have in place an adequate organizational structure and sufficient personnel to assure that the devices the manufacturer produces are manufactured in accordance with the requirements of this regulation. Each manufacturer shall prepare and implement quality assurance procedures adequate to assure that a formally established and documented quality assurance program is performed. Where possible, a designated individual(s) not having direct responsibility for the performance of a manufacturing operation shall be responsible for the quality assurance program.

(a) Quality Assurance Program Requirements.

The quality assurance program shall consist of procedures adequate to assure that the following functions are performed:

(1) Review of Production Records;

(2) Approval or rejection of all components, manufacturing materials, in-process materials, packaging materials, labeling, and finished devices; approval or rejection of devices manufactured, processed, packaged or held under contract by another company;

(3) Identifying, recommending, or providing solutions for quality assurance problems and verifying the implementation of such solutions; and

(4) Assuring that all quality assurance checks are appropriate and adequate for their purpose and are performed correctly.

(b) Audit Procedures.

Planned and periodic audits of the quality assurance program shall be implemented to verify compliance with the quality assurance program. The audits shall be performed in accordance with written procedures by appropriately trained individuals not having direct responsibilities for the matters being audited. Audit results shall be documented in written audit reports, which shall be reviewed by management having responsibility for the matters audited. Follow-up corrective action, including re-audit of deficient matters, shall be taken when indicated. An employee of the Food and Drug Administration, designated by the Food and Drug Administration, shall have access to the written procedures established for the audit. Upon request of such an employee, a responsible official of the manufacturer shall certify in writing that the audits of the quality assurance program required under this paragraph have been performed and documented and that any required corrective action has been taken.

It is the responsibility of each manufacturer to determine the quality program that is capable of achieving the quality objectives. Once the quality program is established, it must be maintained. Quality programs are also an integral part of international standards, such as ISO 9000 standards.

QUALITY ASSURANCE OR QUALITY CONTROL?

QA is a total system approach to guarantee the safety and efficacy of finished products. It goes beyond simple inspection and testing. It must be considered at all stages that have an effect on the ultimate quality, safety, and effectiveness of the product, including development, design evaluation, component or vendor selection, process development, validation, pilot production, manufacturing, testing and inspection, records, labeling, distribution, and complaints. In order to be effective, quality consciousness must be fostered in every employee, from top management down. A QA system is documented and executed in order to achieve the stated quality objectives. The written policies are established by management and practiced by all employees.

Quality control (QC) is usually considered as a subset of the total QA system. It is the most basic or minimum type of a quality program. Typically, a QC program addresses the evaluation of raw materials, in-process materials, packaging, labeling, and finished products in order to prevent production and shipment of defective or substandard products. QC primarily consists of inspection and testing as the primary means of detecting defects. It is only a small part of the total QA program that achieves complete control. An effective QA program will prevent defective products and correct problems leading to any defective products, not merely identify and set aside the defective products. QC testing and inspection is more of a reactionary process rather than a proactive process. QC remains an important part of the QA system when it provides information that is fed back into the system to identify and correct the root causes of any quality problems.

Many people are confused by the terms *quality assurance* and *quality control*. The functional department is termed "quality assurance" in the device cGMPs and the "quality control unit" in the drug cGMPs. Additionally, the term used in the Good Laboratory Practices (GLPs) is the "Quality Assurance Unit" (QAU). The difference between QA and QC is mostly operational. The QAU is usually responsible for performing the tests and inspections to ensure that the specifications are met and the limits are adhered to. The QAU is responsible for auditing methods, results, systems, and processes. QA is responsible for performing trend analysis. The requirements for QA and QC can be accommodated in organizations that have both QA and QC departments, either separate or combined. The U.S. Food and Drug Administration (FDA) is less concerned with the name of the function than with the fact that the function is accomplished.

QUALITY ASSURANCE COMPLIANCE PROGRAMS

A good QA program consists of a number of compliance programs that accomplish the quality objectives effectively. Many of these programs are universal in their application to various product types, while others are unique to particular product types. General QA programs would include an effective program for employee awareness of quality. The system should include a formal training program for manufacturing and QA personnel. Through this training, all employees are made aware of the role of quality practices. While QA has the primary responsibility for QA program management, other departments have important quality functions also, including designing quality into the product, documenting the design, designing quality into the manufacturing process, and producing products that meet the quality standards.

Key quality programs include, but are not be limited to, documentation systems, training, calibration, maintenance, audits, validation, change control systems, complaints, cleaning and sanitization, environmental monitoring, and stability programs.

Documentation

Documentation includes production and control records, Standard Operating Procedures (SOPs), record keeping, numbering systems, distribution records, inventory records, and so on (DeSain 1992). In order to control records and documents effectively, it is critical to establish numbering systems for the records and documents. These numbering systems must be unambiguous, yet flexible enough to allow for a variety of circumstances that may arise in the future. There should be numbering systems for SOPs; production and specification documents; part and lot numbers for components, raw materials, intermediates or work in progress, and final products; identification numbers for equipment; and numbering systems for validation documents, technical reports, and so on. The numbers should be logical, traceable, unique, and controlled. SOPs should be written and followed that describe the numbering system and the personnel responsible for assigning and controlling numbers.

Standard Operating Procedures

SOPs are the primary documents of process control. The first SOP to be written should be a procedure for writing SOPs that defines the format of all ensuing SOPs. This procedure should describe the format and contents of the procedures to be written, as well as the categories of procedures. The responsibility for writing, approving, distributing, controlling, and numbering the documents should be addressed. SOPs should cover administrative functions, QA, maintenance, material handling, production, and QC. Typically, an SOP will include a title, number, purpose/scope or objective, instructions or procedure, and approvals. Additional sections may include references, preliminary operations or requirements, acceptance criteria, definitions, data forms or references to forms, safety precautions, or other subjects as deemed suitable.

Personnel Training

The training of personnel is a significant function that must be performed. Personnel who are given tasks to perform must be qualified to carry out those tasks effectively. They must have an understanding of the regulatory requirements, such as the cGMPs, and they must understand the impact of failure to perform their tasks in compliance with those regulations and standards. In addition, personnel must have proper technical knowledge and experience to perform their assigned functions. The qualifications can come from formal education, on-the-job training, certifications by professional societies, past job experience, or any combination of these. There should be a formal job description for each position, defining what must be done and what training is necessary to qualify the individual for the job. There should be a formal training program that is written and implemented, defining the requirements for education and training for each position. It should also define the documentation that is required for the training records. The individuals who provide the training must be qualified to provide the training. A record of training, which provides the identity and qualifications of the trainer, the topic or contents of the training, the date and time or length of the training session, and the identity of the individuals who attended the training, must be maintained. Two categories of training are general GMP training and specific job or departmental training. Often, the GMP training is accomplished by the QA organization, while the specific job training is performed within the various departments.

Calibration

A calibration program is another critical program specifically required by the cGMPs. An effective calibration program consists of a listing of all equipment to be calibrated, the procedures for the calibration and maintenance of instruments and equipment, and records of calibrations. These procedures define the overall calibration program, as well as the specific procedures for calibrating each instrument. Included in the procedures is the assignment of responsibility for the calibration program, description of records to be kept, frequency of calibration, and determination of calibration expiration dates. Action limits on calibration must be established and defined. There must be reference to traceability to primary standards of the U.S. National Institute of Standards and Technology (NIST) or other acceptable standards. The equipment used to calibrate instruments also needs to be properly calibrated and referenced to NIST traceable standards. Calibration stickers or tags should be on or near the equipment, showing the calibration date and the date the next calibration is due for ready reference by the operator.

Maintenance

Maintenance programs are also essential programs that must be established. A maintenance program is in many ways similar to a calibration program, in that it will include a complete listing of all equipment to be included in the maintenance program. Further, SOPs describing the overall maintenance program, as well as specific procedures for the maintenance of the individual pieces of equipment, should be developed. The maintenance procedures include assignment of responsibility, maintenance schedules, descriptions of the methods and materials used in maintenance, and records of the maintenance activities. Routine preventive maintenance is included, along with provisions for emergency or nonroutine maintenance. All records for the equipment are kept in equipment history files. Maintenance must be performed by properly trained and qualified individuals.

Audits

In order to ensure that all systems and programs are functioning properly, an effective audit program is essential and required. An audit is an independent review of a program, department, or other system to evaluate its conformity to a standard. An obvious standard would be the cGMP regulations, but compliance or conformance to one's own procedures is equally important. An audit will evaluate if written procedures exist, if the procedures are being followed, and if the procedures are effective when they are followed. Audits can be performed internally—various departments or systems are evaluated by an independent audit team, usually including QA personnel—or externally—audits conducted to evaluate the compliance of vendors, contract manufacturers, or test laboratories that are providing a service to the company. An effective audit program will have a list of areas to be audited and an audit schedule and proper standards to audit against. Records of all audits must be maintained.

An important element of effective audit programs is the follow-up to the audit findings. Are corrections made in a timely manner to nonconformances found as a result of the audit? The FDA has a policy regarding access to the results of the audits. The FDA will not review or copy a company's audit records and reports when these audits are performed in accordance with their written QA program on audits. The intent of this policy is to encourage companies to conduct meaningful audits. The main concern of the inspectors is that audits were, in fact, conducted according to the procedures. Therefore, they may review the schedule for audits and the records confirming that the audits did take place and that any required corrective action has been taken.

Validation

Validation has been a focus of attention for a number of years. In order to achieve a high degree of confidence that a product is acceptable and meets its predetermined specifications, the reliability of the process for producing the product must be verified. Validation of the equipment used to make the product, the methods used to test the product or monitor the process, the raw materials and components that are used in the production of the product, the facility in which the product is made, and the personnel who make the product and perform the tests are all part of the total validation process. The FDA guideline on general principles of process validation (FDA 1987a) is a key document to the concept of validation. Other FDA documents address specific validation issues, such as cleaning, assays, and so on.

Change Control

Once a process is established through validation, a procedure is written and approved, or a manufacturing procedure is written and implemented, a program to control changes to that process, procedure, or document must be established to prevent unacceptable or unauthorized changes from occurring. Prior to the change or alteration of any equipment, instrument, system, process, or procedure involved in the manufacturing, testing, and/or handling of a component, drug product closure or container, in-process material, or final product that may impact the safety, quality, strength, purity, or identity of the product, a formal mechanism for the review and acceptance (or rejection) of such change must be established. The change must be reviewed and receive formal authorization. Procedures for the control of changes to documents or procedures must be written and implemented. QA has the responsibility to see that the changes are handled properly. Historical records of changes should also be maintained. This includes not only the changes but also the data and rationale that support these changes.

Environmental Control

To achieve and maintain environmental control, an environmental monitoring program must be developed and implemented. The program is designed to monitor activities that may affect product/process integrity, personnel protection, and the cleaning and sanitization program. Particulate and microbiological monitoring of the equipment, personnel, and environment represents an important feature of any control program for the manufacture of drug products. Controlled areas must be maintained at a reasonable microbiological level to prevent the presence of excessive microorganisms that may reflect inadequate cleaning, sanitization, and disinfection practices, especially in areas where sterile products are processed. Nonviable particulate levels must be monitored to ensure that cleaning and air handling are being performed adequately. Routine monitoring is considered as appropriate to demonstrate adequacy of environmental control conditions. Programs for environmental monitoring must be written and implemented.

Facility and Equipment Cleaning

Equipment and facilities must be maintained in a clean and orderly manner. In recent years, there has been a major emphasis placed on cleaning and cleaning validation. The FDA has issued a guide for the inspection of cleaning validations (FDA 1993c), which addresses specific issues on the approaches to cleaning validations. The guide was issued in response to incidences that raised the awareness of the FDA to the potential of cross-contamination due to inadequate cleaning procedures. One particular incident was the contamination of a bulk drug pharmaceutical chemical by low levels of intermediates and degradants from the production of agricultural pesticides. The containers that were used to store recovered solvents had previously been used to store pesticides, but the containers had not been adequately cleaned. Practices that the FDA has found to be acceptable and some that are unacceptable are discussed in the guide. The FDA expects companies to have written procedures describing in detail the cleaning processes for various pieces of equipment.

There should be general written procedures that define how the cleaning processes will be validated. Included in the procedure should be the person(s) responsible for performing and approving the validation study, the acceptance criteria, and when revalidation will be required. Specific written and approved validation protocols are to be prepared in advance for studies that address sampling procedures, analytical methods, and the sensitivity of analytical methods. The studies are expected to follow the protocols, and the results of the studies are to be documented, along with conclusions arrived at from the studies. A final validation report must be prepared and approved by management. The report should include a statement as to whether or not the cleaning process is valid.

An important issue to be considered in the cleaning of equipment is the design of the equipment for cleanability, especially in clean-in-place (CIP) systems. Design considerations might include the type of valves used in a system and whether or not they are truly cleanable by a CIP process. Operators must be trained to perform the cleaning process properly. Critical factors to be considered are the length of time between the use of the equipment and the cleaning and between cleaning and the subsequent use of the equipment. Dirty equipment that is allowed to stand for a long period of time before cleaning can have dried-on protein or other types of contamination that is much more difficult to remove than if it is cleaned immediately after use.

All cleaning operations should be documented. For more complex cleaning procedures, the documentation may be more detailed than for a simple cleaning process. Information that should be included is the identity of the operator(s) performing the cleaning, the date and time of the cleaning, the procedure followed, and the product that was in the equipment in the previous batch. The analytical methods used in the validation of the cleaning process must be understood in terms of sensitivity and specificity. Limits for acceptability of the cleaning process must be established and justified. If detergents are used in the cleaning process, the validation must include provisions for determining their residual levels. The concept of routine cleaning and recleaning until an acceptable result is obtained is not acceptable. Constant testing, resampling, and retesting is evidence that the cleaning process is not validated and is ineffective.

SPECIAL CONCERNS FOR BIOTECHNOLOGY PRODUCTS

Although products produced by biotechnology are governed by the same rules and regulations as classical drug products, there are often specific regulations or standards that have been developed for biotechnology-derived products. Characterization of the products to ensure safety, purity, and activity or efficacy incorporates not only classical methods such as USP monograph methods (USP 23/NF 18 1995) but also methods that are specific to the particular technology used for the production. Biological products are likely to cause unwanted effects, such as immunological sensitization and response in patients. Slight modifications in the product can enhance these effects; therefore, it is desirable to characterize the products carefully.

Some of the methods used for biotechnology-derived products that are the same as those used for traditional pharmaceutical products include product sterility, safety in experimental animals, and potency. The use of reference standards and validated methods to assay for contaminants and impurities is similar. The fundamental difference between the QC systems for biotechnology-derived products and other pharmaceutical products is in the methods that are used to determine the product's characteristics: identity, purity, and impurity profile. Often, a combination of final test procedures, in-process testing, and process validation is necessary. The complexity of this combination may relate to the production process being used. Part of the QC of cell-derived products is the complete characterization of the cell line, going all the way back to the origin of the cell line. The characterization may include testing for adventitious agents, phenotyping, and antibiotic resistance. QC of the original master cell bank (MCB) and the working cell bank (WCB) may include identity and stability monitoring, as well as testing for adventitious agents, retroviruses, retroviral activity markers, and tumorgenicity.

Biotechnology Test Methods

The analysis of biotechnology-derived products involves the use of sophisticated analytical methods. The technology is constantly changing. There are a number of specific analytical methods that are particularly useful for demonstrating the structural identity and homogeneity of these products. These methods are used to determine the stability of the products. A discussion of some of these special methods follows.

Protein Analysis

Protein assays are used for the quantitative determination of total protein content in a preparation. What may seem to be a relatively simple task often turns out to be a complex and difficult measurement. Complications include contamination of the material of interest by other proteins, unknown or inaccurate absorption coefficients, a lack of reference standards, and nonspecificity of the assays. Frequently, one method is not sufficient for the protein determination, and additional methods must be used to confirm the values obtained.

Absolute determinations of protein content sometimes can be performed independent of reference standards. Ultraviolet (UV) spectrophotometry can be used without a reference sample if there is a known, valid absorptivity for the material of interest. UV absorbance is best used with very pure proteins, since other proteins may interfere with the test results. This simple method is based on the UV absorption of the protein in solution. The absorbance at the absorption maximum wavelength is determined, and the protein concentration is calculated using an empirical extinction coefficient. This method is useful for proteins that contain aromatic amino acid residues. UV determinations are subject to interference by large aggregates that cause light scattering and contaminants that also have UV absorption at the wavelength being used.

Kjeldahl nitrogen analysis is another method that can be used independent of a reference standard. This method gives a precise and accurate determination of total protein. The protein sample is decomposed in sulfuric acid, and a determination of the ammonia produced is accurately measured. Factors are used to calculate the protein concentration.

The Lowry method is one of the most common methods of determining total protein content. It is based on the biuret reaction of proteins with copper in a basic solution and subsequent reduction of the reagent to produce a blue color. The blue reaction products are measured spectrophotometrically to give the protein concentration by comparison to a reference standard.

Other methods that require the use of reference standards also can give accurate and reliable results. These methods include biuret and quantitative amino acid analysis. An additional method, the

Bradford method, incorporates the use of a protein binding dye, Coumassie Brilliant Blue, in an acidic environment. The bicinchonnic acid (BCA) assay is another commonly used protein assay that is less sensitive to impurities than the Lowry assay. Fluorescent methods with fluorescamine or O-phthaldehyde (OPA) can be used because they have the advantage of increased sensitivity. Both react with primary amines at the N-terminus of a polypeptide and with amino acid side chains.

Amino acid analysis is more complex, but it can be used to either determine quantitatively the protein concentration or the actual absorptivity of the protein. It is a classical protein chemistry method to determine the amino acid content of a protein or peptide. The protein is hydrolyzed to its component amino acids, and these are then separated and quantitated chromatographically. Amino acid analysis can be used to determine both the amino acid composition and the total amount of protein.

Protein Sequencing

Protein sequencing is a step beyond amino acid analysis and can be used to characterize the protein molecule. It can provide primary structure information and is also used to confirm protein homogeneity. There are two types of protein sequencing: amino-terminal (N-terminal) and carboxy-terminal (C-terminal) sequencing.

N-terminal sequencing is based on classical protein chemistry methods reacting with the amino-terminal residue of a protein with phenylisothiocyanate (PITC) and the cleavage of the derivative from the protein with a perfluoridated acid. This cleavage exposes the next amino acid, which serves as the new N-terminus for subsequent coupling and cleavage cycles. There are a number of automated protein sequencers available commercially. The procedure can also be performed manually. The analysis of the cleavage products by reverse phase high performance liquid chromatography (HPLC) provides information that can be used to determine the sequence and quantitation of the amino acids.

C-terminal sequencing gives information about the primary structure of the protein. The sequential degradation of the protein can be done either by enzymatic or chemical methods. The protein digests can be analyzed by HPLC, mass spectrometry, or a combination of these two methods.

Peptide Mapping

Peptide mapping is a highly specific identity method. It can be used as a tool for confirmation of genetic stability and to compare the protein structure of a specific lot of product to a reference standard or to previous lots of the product to confirm the primary structure and to show conformance to lot-to-lot consistency. Peptide mapping involves selective fragmentation of the protein into discrete units that are resolved by chromatographic techniques. Fragmentation is performed either with a protease, such as pepsin, chymotrypsin, or other enzymes, or with selective chemical degradation. Analysis of the resultant degradation products is performed with a chromatographic method such as reverse phase HPLC and/or high performance ion-exchange chromatography (HPIEC).

Immunoassays

Because of the sensitivity and specificity of immunoassays, they can be used as identity methods for active drug substances or as quantitative methods to determine either the protein of interest or impurities. Since many products are prepared in *Escherichia coli* (*E. coli*) or Chinese hamster ovary (CHO) cells, sensitive immunoassays for *E. coli* and CHO proteins have been developed that can measure these proteins to very low levels. Immunoassays also can be used as potency assays for monoclonal antibodies using the appropriate antigens.

Immunoassays depend on specific high-affinity antibody: antigen interactions. Examples of immunoassay methods include radioimmunoassays (RIAs), enzyme-linked immunosorbent assays (ELISAs), and immunoradiometric assays (IRMAs). The development of an immunoassay for a biotechnology-derived product involves the production of the antisera or antibody, the preparation of a labeled tracer, and the preparation of reference standards. A means of separation of the free antigen from the bound antigen must be developed. The ELISA format is often a sandwich assay that utilizes two antibody preparations, like IRMA, but with an enzyme label rather than a radioactive label. Typically, the enzyme label is horseradish peroxidase (HRP) or alkaline phosphatase.

Electrophoresis

Electrophoretic assays are powerful assays that can be used to determine protein purity and homogeneity. The methods also are good as stability-indicating assays to identify any molecular or chemical changes, such as degradation, aggregation, oxidation, deamidation, and so on. The methods are simple and require small samples of product. The most common electrophoretic methods are sodium dodecyl sulfate–polyacrylamide gel electrophoresis (SDS–PAGE) and isoelectric focusing (IEF).

SDS–PAGE separates proteins by molecular weight. The sample is first denatured in the presence of the anionic detergent (SDS), then the complex is electrophoresed through the polyacrylamide gel support. Protein migration through the gel support is proportional to protein size—small proteins migrate faster than larger ones. Samples are often electrophoresed in both reduced and nonreduced states to determine if proteins of the same molecular weight or intramolecular proteolytic cleavages of the proteins of interest are present. SDS–PAGE gels can be used to determine if there are aggregations or oligomerization of the protein of interest, but only if the aggregates or oligomers are stable in the presence of SDS. Methods to detect the sample after electrophoresis include densitometric analysis with Coumassie Brilliant Blue stain or silver stain. Silver stains can be performed quantitatively under proper conditions. SDS–PAGE with Coumassie Brilliant Blue is used to determine the purity of the sample quantitatively. SDS–PAGE separation can also be combined with immunological methods such as immunoblotting. This is known as western blotting, which can be used to determine the identity of the protein band in question.

In the IEF method, proteins are separated on the basis of their charge in an electric field. For each protein, there is a pH at which the protein is isoelectric, resulting in the various charges on the protein canceling each other out, thus making the net charge essentially zero. IEF is done in the native state in a support of loose pore polyacrylamide or agarose gel containing amphoteric, low molecular weight ions that set up a pH gradient because of their migration within the support matrix when an electric field is applied. In the presence of an electric charge, positively charged proteins migrate toward the cathode, and negatively charged proteins migrate toward the anode. Migration stops when the protein reaches the pH value in the support gradient where the net charge becomes zero (neutral). The migration is dependent on the amino acid composition; therefore, the altered forms of the protein or other proteins migrate to different points on the support. IEF gels are stained with Coumassie Brilliant Blue or silver stains. IEF can be used as a tool for identifying or ensuring the homogeneity of a protein. It is also a good tool for determining the stability of a protein. The extent of glycosylation can be determined for a monoclonal antibody using IEF.

FINAL PRODUCT LOT RELEASE

Biological products have singular issues for lot release in addition to those for more conventional pharmaceuticals. The FDA has provided guidance for the development, manufacture, and safety evaluation for biological products in the various "Points to Consider" documents issued by the Center for Biologics Evaluation and Research (CBER) (FDA 1985, 1987c, 1994). There are similar documents available that address biological products intended for licensing in the European Community (EC), Canada, and Japan.

In order to release a lot of biological product, data must be submitted to the regulatory agencies to demonstrate that the product meets critical biological criteria. Reviews (Rhodes 1985; Schiff et al. 1992) have been published that provide an overview of final product testing, with an emphasis on the various safety tests that detect biological contamination in various products. Critical product criteria for release and characterization of biological products are safety, potency, purity, and efficacy. The safety tests include general safety, pyrogenicity, sterility, mycoplasma contamination, DNA (deoxyribonucleic acid), endogenous retroviruses, and exogenous viruses.

General Safety Testing

The general safety test is described in 21 CFR 610.11 (Code of Federal Regulations 1995) and in the USP (<88>, pp. 1702–1703). (It is also defined by the EC as "Abnormal Toxicity" in the *European Pharmacopoeia* [EP], part 1, V.2.1.5). This test is designed to detect any general unexpected and unacceptable toxicity. It is a seven-day test using guinea pigs and mice that are inoculated with prescribed amounts of the product. The animals are observed for the seven-day period. The product passes the general safety test if all inoculated animals survive the test, show no unexpected responses, and no weight loss is observed during the test. There is a provision for repeat testing in the event of a failure on the initial test. An additional number of animals is used for the retest, and the repeat is satisfactory only if all of the retest animals survive the test period without showing any unexpected response.

Sterility Testing

The sterility test is described in 21 CFR 610.12 (Code of Federal Regulations 1995) and in the USP (<71>, pp. 1686–1690). For biological products, both the final bulk and the final filled material must be inoculated into fluid thioglycollate broth, incubated at 31°C, and in soybean-casein digest broth, incubated at 23°C. To detect fungi and anaerobes, inoculation into Sabouraud-dextrose broth, incubated at 23°C and 36°C, and peptone yeast glucose broth, incubated anaerobically at 36°C, are also recommended. The actual volumes of product used for the test are determined by the number and fill size of the final product. For a product to pass the sterility test, no growth may be observed in any of the product-inoculated cultures. Although the USP describes a retest option if the test is unsatisfactory (i.e., growth is observed), it is not generally acceptable to release a product on the basis of the retest.

The membrane filtration method is another method described in the USP for the determination of sterility. In this test method, a prescribed volume of the product is filtered through a membrane filter (0.45 μm), and the membrane filter is added to either soybean-casein digest medium at 20–25°C or thioglycollate medium at 30–35°C. The media are observed for evidence of growth for 7 days. The details of the procedure are slightly different for products intended for intravenous administration than for other types of administration. Prior to running the assay, the level of bacteriostatic and fungistatic activity of the product must be determined. The product is considered acceptable if no growth is observed. Even if no growth is observed, but a review of the procedures and facilities demonstrate inadequate or erroneous performance of the test, the test can be declared invalid and repeated. Again, if the test fails because growth is seen, a retest procedure is described. Even though both of these sterility tests technically allow for retests, there is a tendency to not allow the retests as valid results unless there is documented evidence of a test procedure failure.

Pyrogen Testing

Pyrogen tests are methods to determine the ability of the final product to cause fever as a response to its administration. The rabbit pyrogen test, as described in the USP (<85>, pp. 1696–1697), is an acceptable method for testing the final product. A prescribed dose of the product is administered to 3 rabbits, and the body temperature of the rabbits is monitored rectally for 3 h. The product is determined acceptable if no rabbit shows a rise in temperature of $\geq 0.6°C$, and the total temperature rise of the 3 rabbits is $< 1.4°C$. If the product fails the initial test, there is a provision for a retest using an additional 5 rabbits. The product is considered to pass the retest if no more than 3 of the 8 rabbits (3 original and 5 retest) show individual temperature rises of $\geq 0.6°C$, and the sum of the 8 individual temperature increases does not exceed 3.7°C.

An alternative to the rabbit pyrogen test is the Limulus amebocyte lysate (LAL) assay. The LAL test actually determines the level of endotoxin, which is the most common pyrogenic substance encountered. In order to use the LAL test for final product release, the assay must be validated in the testing lab according to the guidelines issued by the FDA (FDA 1987b). Validation of the assay includes qualification of the laboratory and inhibition and enhancement tests on at least three production batches of each finished product. There are a number of commercial LAL tests on the market, including gel clot, colorimetric, turbidimetric, and kinetic assays. The product is considered acceptable when the endotoxin levels are below FDA guideline levels. If the company obtains approval to use the LAL test for final product release, it is not acceptable to switch back to a rabbit pyrogen test in the event the LAL test fails.

Mycoplasma Testing

The "Points to Consider" and 21 CFR 610.30 (Code of Federal Regulations 1995b) require the product to be free of mycoplasma. The European market references the EP method, which is similar to the CFR methods, except for using different positive controls. In order to detect low levels of mycoplasma of various types adequately, two methods must be used: direct and indirect determinations. Agar plates and broth medium are inoculated aerobically and anaerobically in the direct assay. The semisolid broth cultures are subcultured on the 3rd, 7th, and 14th days, and again incubated both aerobically and anaerobically. The culture plates are incubated for no less than 14 days at 36°C and observed microscopically for growth of mycoplasma colonies. Positive controls are prescribed in the methods defined by the FDA. The indirect test uses a Vero cell culture substrate inoculated with the product. The cell cultures are incubated for 3 to 5 days at 36°C in a 5 percent carbon dioxide atmosphere. The cultures are then examined for the presence of mycoplasma using epifluorescence microscopy and a DNA binding fluorochrome stain. The product passes the mycoplasma test if no broth, agar, or Vero cell culture shows evidence of mycoplasma.

DNA Testing

DNA contamination is determined most commonly by the hybridization analysis using species-specific probes. The method includes a spike and recovery procedure that ensures an efficient quantitative extraction of DNA from the sample, even in the presence of high protein concentrations. The alternative to the hybridization test for DNA is the Threshold Total DNA Assay (Molecular Devices, Menlo Park, Calif., USA). This method is based on the capture of DNA onto a membrane, and the amount of DNA can be measured by changes in surface potential on an electronic chip. This procedure is not species specific and finds its greatest utility in validation studies performed to document the removal of DNA in spike and recovery studies.

Virus Testing

The possibility that viable viruses might remain in purified products exists. Such viruses could infect humans. The types of viruses that must be tested for varies according to the method of preparation of the product. MAb products prepared from mouse cells may contain host mouse viruses as well as viruses propagated by hybridoma cells. Virus testing requirements have been reviewed by Joner and Christiansen (1988). There are a number of methods that can be used to determine the presence of viruses. Some of these assays are specific, while other methods are quite general.

The tissue culture safety test is a general screen for a wide variety of infectious viruses and uses isolation assays on several sensitive cell lines. These cell lines will exhibit reactions to several viruses, including herpes viruses, paramyxoviruses, parainfluenza viruses, poxviruses, and rhabdoviruses. Several assay methods are used to determine the presence of and contamination by other specific viruses. Hepatitis B virus (HBV) can be detected by a number of third generation test kits that are commercially available. For the Epstein-Barr virus (EBV), viral DNA hybridization, nuclear antigen fluorescent staining, and immortalization of umbilical cord blood lymphocytes are suggested screening procedures. Cytomegalovirus can be detected by tissue culture isolation techniques. Retroviruses can be detected by biochemical enzyme assays, electron microscopy, and DNA hybridization. There are a number of murine viruses of concern that can be effectively tested for by the mouse antibody production (MAP) test. Typically, mice are injected with the product of interest through different routes and then maintained for a period of 28 days. Serological tests then determine whether the animals have any detectable antibodies to the viruses in question. Additional specific virus tests include that for lactic dehydrogenase elevating virus (LDHV), in which the animals are challenged with a viable virus. The surviving animals are assumed to have built up an immune response to the virus, if it is present in the test sample. LDHV is determined by assaying for increases in lactic dehydrogenase levels in the animal's serum 3–7 days after the injection.

Concluding Comments

The safety of the product is evaluated on more than just the final product. The safety evaluation will include a number of steps leading up to the final production. First, there is an initial screen of the candidate cell line. The characterization of the MCB and the manufacturers working cell bank (MWCB) follows. Testing continues into the in-process testing stages of bulk harvest fluid and postproduction cells. The purification process must be validated for the removal and/or inactivation of viruses, DNA, and mycoplasma. The process is also examined for other routes of contamination. The main routes of contamination are from raw materials, operators, equipment, and the process itself. The complete validation will consider all of these avenues as potentials for contamination of the final product with adventitious agents.

PROCESS VALIDATION

Once the manufacturer has identified a target process, the validation exercise begins. The critical process steps are identified and selected for study via a series of prevalidation or development studies. The process parameters are evaluated, and acceptable ranges of variation are established. These ranges should include the expected operational limits that would be encountered in actual production. The specifications for the resulting product or intermediate are established from prevalidation or development studies. The test methods and challenges to the process are determined. Acceptance criteria are agreed to by a validation team consisting of appropriate personnel from various functional areas, which may include QA/QC, manufacturing, R&D, technical support, regulatory affairs, and so on. The validation protocol is developed and approved by individuals representing the functional groups from the above list prior to beginning the execution of the protocol. A number of prevalidation tasks are also completed or verified before the protocol is started. These tasks may include, but are not limited to, equipment calibration, certification of laminar flow work areas, installation qualification (IQ) and operational qualification (OQ) of any equipment, the release of raw materials, training of individuals, and validation of assay methods.

The protocol is executed with generally three or more (in the United States) or five or more (in Europe) runs completed for the validation. The actual process parameters and test results are collected and compared to the predetermined requirements. Additional data are sometimes recorded that may not have predetermined requirements in order to enhance the understanding of the process. A complete validation package is assembled, including copies of original documents, SOPs used during the study, manufacturing and test equipment logs, parts lists, process reports, and other relevant validation reports.

Process Parameters

Process parameters are not the only parameters to be considered in the course of a validation activity. Controllable process parameters such as temperature, pH, time, rate of addition and/or mixing, and so on are included in the validation. The influence of these parameters should be characterized during development studies; those parameters that have a significant effect on the resultant product must be included in the validation studies. Resultant process parameters must also be observed and evaluated. Some key process parameters include the following:

- **Yield:** Although yields are more a product of the system rather than a controllable parameter, it is important that expected ranges be established for each step in the process. A sudden departure from the expected yield for the step may signal the advent of serious process difficulties.

- **Time:** Time intervals between processing steps can be a source of problems. The elapsed time of a process is a critical parameter that should be monitored and controlled. The time between processing steps and the time of a process step should both be included as part of the validation study; for biological products, however, these may be especially long. A process that is under control should have stable processing times for individual steps and for the overall process. It is important to execute the process steps within the time frames and at the limits of the time frames that will be allowed during the processing operation.

- **Purity:** As with yields, the purity of the product of the process step is a process parameter that is not necessarily controllable. Expected ranges of purity for each processing step should be established. Purification factors from a particular step are also an important parameter because it serves to monitor the operational effect of the process step.

In the event of a deviation from the protocol or the expected results, or if a clarification is needed to enhance the understanding of a particular report, comment/deviation reports and subsequent investigation reports are to be completed, approved, and included with the validation package. A summary report is written to describe the results, analyses, and conclusions of the total validation package. The summary report and the complete data package are reviewed and approved. The cover summary report and the complete validation package are archived for future reference.

Cell Line Characterization

Products produced by a cell line as part of the process, for example, monoclonal antibodies produced by hybridoma cell lines made by chemically induced fusion of the antibody-producing cells with myeloma cells, will need to be validated. As part of the validation of the process, the cell lines will be documented for the following information:

- The source, name, and characterization of the parent myeloma cell line, including any immunoglobulin heavy or light chains that it secretes.

- The species, animal strain, characterization, and tissue origin of the immune cell.

- Description of cell-immortalization procedures used in generating the cell line.

- Identification and characterization of the immunogen.

- Description of the immunization scheme.

- Description of the screening procedure used.

- Description of the cell cloning procedures.

- Description of the seed lot system for establishing and maintaining the MCB and the MWCB.

Virus Removal and Inactivation

The nature of production from tissue culture methodologies makes the product susceptible to viral contamination. Viral contamination can come from two sources: endogenous viruses that are covalently linked to the host genome and adventitious viruses that are external to the host genome. It is necessary that the MCB and the culture medium be free from oncogenic contamination. The characterization of the MCB and the MWCB will make that determination.

The recommended procedures in the EEC guidelines (EEC Ad Hoc Working Party 1991) can be used as the guide for virus clearance and inactivation studies. These guidelines require that the cell line characterization of the MCB and the MWCB be documented. It must also be demonstrated that viruses are not copurified with the product; hence, viral clearance and viral inactivation studies will be performed as part of the validation. The removal and/or inactivation will be performed with model viruses. The product will be spiked with model viruses that can be cultivated at high titers. Several model viruses will be selected for the studies, encompassing large and small particles, DNA and RNA (ribonucleic acid) genomes, as well as enveloped and nonenveloped strains. Both chemically sensitive and chemically resistant strains should also be considered. A determination of viral burden should be performed on product prior to purification, which may impact the actual selection of model viruses.

Murine cells used to produce monoclonal antibodies are inherently capable of producing infectious murine retrovirus. Therefore, it is not necessary to confirm the presence of murine retrovirus; rather, the amount of retrovirus in the unprocessed bulk should be quantitated in a series of bulk harvests to show consistency from lot to lot.

At least one or more steps that are known to remove or inactivate viruses should be included in the purification process. Validation is performed by spiking the product with the model viruses, and recovery studies for the viruses will be performed before and after each step. Each purification step will be evaluated for the removal or inactivation of viral contamination.

By determining the viral removal or inactivation for each step, the total viral reduction capability of the purification process is determined by adding together the log reductions for each step. Only steps that result in two or more log reductions will be considered in the total capability of the process. Identical steps that are repeated will only be considered once in determining the total capability of the process. Spiking experiments can be performed at a small scale, simulating the purification step as closely as possible.

DNA and Nucleic Acid Removal

Potential sources of nucleic acid are from the host cell DNA and retroviral RNA. The FDA *Points to Consider* and the World Health Organization (WHO) have set limits on DNA contamination. In a manner similar to viral inactivation and removal validation, the removal of nucleic acids in the purification process is demonstrated by clearance experiments with spiked DNA.

Each purification step should be evaluated for DNA clearance. The size of the DNA used for spiking should be similar to that found in the host. The total clearance factor can be determined as the sum of the clearance factors for each step. In the same way as in viral clearance studies, the spiking experiments for DNA removal may be performed at a small scale, simulating the purification step as closely as possible.

Other Contaminant Removal

Other potential contaminants include transferrin, insulin, serum albumin, and related proteins, such as fragments, aggregates, and/or denatured product. Additional contaminants may arise from the purification process itself and may include resin fragments and ligands, buffer salts, or components. The list of all known potential contaminants will include physical properties of the potential contaminants, such as molecular weight, isoelectric point (pI), and so on.

Potential contaminants are to be identified and characterized. Purification steps to remove the contaminants should be evaluated for the effectiveness of removing the contaminant. Assays for the contaminants will be validated and documented, including such characteristics as limits of detection.

INSPECTIONS: FDA INSPECTION GUIDES

The FDA has issued a number of documents that are collectively known as inspection guides. These guides are used for the training and educating of FDA inspectors so that they can provide more consistent and uniform inspections. Although a number of these guides can be cited, the two that are particularly germane to this chapter are those for the inspection of pharmaceutical QC laboratories (FDA 1993a) and the inspection of pharmaceutical microbiological QC laboratories (FDA 1993b).

The FDA has recognized that a significant portion of the cGMP regulations pertain to the QC laboratory and product testing. At a minimum, each pharmaceutical QC laboratory should receive a comprehensive review every two years as part of routine FDA inspections. The scope of these inspections may include the following:

- The specific methodology that will be used to test a new product.

- A complete assessment of the laboratory's conformance with the cGMPs.

- A specific aspect of laboratory operations.

Inspections are conducted via a team approach. Although local or district FDA managers make the final decision in assigning the personnel to inspection teams, the team is expected to request additional specific expertise whenever necessary. In preparing for an inspection, the team members review a number of relevant documents, including relevant sections of New Drug Applications (NDAs) or Abbreviated New Drug Applications (ANDAs) or other submissions; FDA publications, such as the Compliance Program for Preapproval Inspections/Investigations; any letters issued by the FDA that describe any previously noted deficiencies that should have been corrected prior to the inspection; replies to the aforementioned letters from the company describing commitments to correct these deficiencies; and the results of previous inspections, including any 483s issued and/or Establishment Inspection Reports (EIRs). It is easy to see that inspectors normally come well prepared for the actual inspection.

Whenever a comprehensive inspection is to be conducted, all aspects of laboratory operations can and will be inspected. Laboratory logs and records will be reviewed for a complete overview of the overall technical ability of the staff and QC procedures. The SOPs should be complete and adequate for all operations, and operators should be following the procedures. Specifications and analytical procedures should be appropriate and, if applicable, in conformance with commitments made in submissions and compendial requirements. The inspectors are instructed to evaluate raw laboratory data; laboratory procedures and methods; laboratory equipment, including maintenance and calibration records; assay method validation protocols; and validation data to determine the overall quality of the laboratory operation and its ability to conform to cGMP regulations. Raw data such as chromatograms and spectra are inspected and evaluated for evidence of impurities, poor techniques, or lack of instrument calibration.

Documentation Review

Even though documents relating to the production of products are reviewed as part of the submissions, the evaluation of these documents relies on accurate and authentic data that truly represents the product. The inspection of the QC laboratory will determine if the data submitted in such applications is, in fact, true and authentic. In many applications, sponsors often do not file all of the test data, since this could create an inordinately large submission and might include redundant data. The inspection team realizes that a company may deliberately or unintentionally select data to submit that shows that the product is safe and effective. The purpose of the inspection team is to determine if there is valid and scientific justification for the failure to report data, should that occur, which demonstrates that the product failed to meet its predetermined specifications. During the inspection, data on batches that were reported should be compared to data on other batches that have been produced. Any exceptions to the procedures or equipment actually used from those listed in the application should be noted.

Out-of-Specification Results

A major concern recently for the FDA is the manner in which a company treats out-of-specification (OOS) results. There should be a system or procedure to investigate laboratory failures. The investigations are a key factor in determining whether a product should be rejected or released, and what the retesting and resampling plan will be. There is a distinction between a product failure and an OOS laboratory result. Laboratory errors occur when mistakes are made by the analyst in performing the procedure, interpreting the results, or using the wrong standards. The occurrence of laboratory errors must be determined by a thorough investigation, not merely assigned arbitrarily as the cause of the failure. The investigation must be documented and consists of more than a retest. There should be a written SOP describing the conduct of the investigation, including the assignment of responsibility for the conduct and review of the investigation. Distinctions should be made between an investigation for a single OOS result and multiple or repetitive OOS results. The following steps are expected to occur for a single OOS occurrence:

- The analyst conducting the test should report the OOS result to the supervisor.

- The analyst and supervisor should conduct an informal investigation that addresses the test procedure, the calculation, the instruments used for the test, and the notebooks containing the OOS result.

The rejection of a test result as an "outlier" must be very carefully considered. Under certain conditions, such as a statistically based test (e.g., content uniformity), it is never appropriate to utilize outlier tests. Companies should not use the outlier concept to reject results on a frequent basis.

Investigation of Errors

If there are multiple OOS results, a more intensive investigation must take place. This full-scale inquiry involves QC and QA personnel in addition to laboratory technicians to identify the errors. If the investigation is inconclusive, the laboratory is not allowed to

- conduct two retests and base release on averaging the results of the three tests,

- use outlier tests for chemical tests, and

- use a resample to assume a sampling error or preparation occurred.

Formal investigations that are conducted should observe the following procedures:

- State the reason for the investigation.

- Summarize the process sequences that may have caused the problem.

- Outline corrective actions necessary to save the batch and prevent a similar recurrence.

- List other batches that might possibly be affected, the results of the investigations of these products, and any corrective action. (This might include other batches made by the errant operator or equipment or the errant process or operation step.)

- Preserve the comments and signatures of all personnel involved in the investigation and the approval of any reprocessed material after retesting.

Any errors, such as mistakes in calculations, should be documented and supported by evidence. Investigations and conclusions must be preserved along with any corrective action. An OOS laboratory result can be invalidated when a laboratory error is documented. However, even if the problem is clearly identified and documented, errors that result from operators making mistakes, faulty equipment, equipment malfunctions, or a deficient manufacturing process are considered product failures.

Any retesting plan should be scientifically sound. The goal of retesting is to isolate OOS results, but retesting should not continue ad infinitum. A product cannot be "tested until it passes." Averaging test results should generally be avoided because it hides the variability of the test procedure. Averaging is particularly a concern when the testing produces both OOS and passing results that, when averaged, are within specifications. Relying on the average figure is misleading and unacceptable.

Laboratory Records and Documents

Laboratory records and documentation are subject to inspection. The analytical notebooks kept by analysts may be reviewed and compared with worksheets submitted and any general lab notebooks and lab records. The records are examined for accuracy and authenticity and to verify that raw data are retained to support the conclusions. Lab records that may be reviewed include equipment logs to confirm the sequence of analyses compared to the sequence of manufacturing dates. Data should be maintained in bound books or on analytical sheets for which there is accountability, such as prenumbered sheets. Records such as laboratory logs, worksheets, or any other written record will be inspected for missing data, data that has been rewritten or corrected, or data that has been concealed (i.e., by correction fluid). Results should not be corrected, modified, or changed without a written explanation. Test results should not be transcribed without retention of the original records, nor should test results be recorded selectively.

Laboratory Standards

The laboratory standards used in the assays must be suitable–labeled, stored properly, dated, and in-date (i.e., not beyond their expiration date). Stock solutions that are to be stored in a refrigerator should be checked to make sure that they are being stored in a refrigerator and that they are properly identified. The preparation of standard solutions should be documented completely and accurately.

Testing Methods

The methods that are used for testing should be validated and documented. If methods appear in the USP, they are considered validated, but the laboratory should still verify that the methods work in their own laboratories. The assay performance characteristics listed in USP 23 Chapter <1225>–"Validation of Compendial Methods"–provide a guide for determining the analytical parameters to be considered in method validation. These parameters include accuracy, precision, linearity, specificity, sensitivity, limits of detection, ruggedness, and others.

In-process testing is also subject to inspection. These tests may actually be performed in the production area, but they must be performed in accordance with the cGMPs.

Laboratory Equipment

The records for using laboratory equipment are subject to inspection. These records include maintenance and calibration logs, repair records, and maintenance SOPs. An important point made in the guideline is that the existence of the equipment specified in the analytical methods should be confirmed, and its condition noted. Inspectors are to verify that the equipment was present and in good working order at the time the NDA (or ANDA) batches were analyzed and that the equipment was being used properly. Inspectors are also to verify that the equipment was in good working order when it was listed as used to produce clinical or biobatches. The concern is that data are suspect if it is generated from a piece of equipment that is known to be defective. To use and release product on the basis of such equipment is a violation of the cGMPs.

Computerized Data Acquisition

Computerized laboratory data acquisition systems are also subject to inspection. This is not new, since a number of compliance guides have addressed this issue. Some key elements that may be inspected include security and authenticity issues. The FDA has provided basic guidance on such issues. Provisions are needed so that only authorized individuals can make data entries. Data entries may not be deleted, and changes must be made in the form of amendments. The database must be as tamperproof as possible. There must be SOPs that describe the procedures for ensuring data validity. The data acquisition system must be validated. One aspect of that validation is to compare data from a specific instrument with the same data electronically transmitted through the system and printed on a printer. This must be repeated over a period of time and with enough frequency that there is a high degree of assurance that the computerized system produces consistent and valid results. The authority to delete files or override computer systems should be evaluated as part of the inspection.

Other Considerations

Inspectors are advised to observe laboratory operations that are actually taking place. There is no substitute for actually seeing the work being performed and observing if good techniques are being used. Part of overall cGMP compliance is to have a documented and ongoing training program.

Microbiological Quality Control Laboratory Inspections

Whereas the *Guide to Inspection of Pharmaceutical Quality Control Laboratories* provides very little guidance on the inspection of microbiology laboratories, the *Guide to the Inspection of Pharmaceutical Microbiology Quality Control Laboratories* focuses on the special aspects of the microbiology analytical process. Each company is expected to develop microbiological specifications for their nonsterile products. In some cases, a company must develop their own special tests, which may not be part of the USP chapter on microbiological limits (Chapter <61>). In this guide, the FDA position on sterility failure retests is stated. They acknowledge that the USP sterility test provides for retests; however, there is a current proposal to remove the retest provision. Regardless, any failure should be reviewed and thoroughly investigated. The rationale is that microbiological contamination may not be uniformly distributed in the lot or batch. Finding a contaminant in one sample and not in another does not discount the results of the first assay. Any retest results should be reviewed, and emphasis should be placed on the logic and scientific rationale for conducting the retest.

Sterility test methods are to be inspected to be sure that procedures are in place to inactivate any preservatives that might be present in the product. The choice of neutralizing agents is dependent on the preservative and the product formulation. The requirements to identify isolates from total plate count testing depends on the type of product. The requirement for an oral solid dosage form may not be as strict as for topicals, inhalants, nasal preparations, and, certainly, injectables.

As part of the inspection, the inspection team is advised to inspect the facilities, equipment, and media. They are warned to be particularly alert to retests or tests that are labeled as "special projects," as these are often studies undertaken on products where contamination has been a problem. The inspectors will review ongoing analyses, such as the previous day's plates and media, and compare what is observed to what is recorded in the logs.

The sterilization process for the media is to be inspected. Problems that are looked for include an autoclave's capability of appropriately displacing steam with sterile filtered air. Autoclaving for too short a time may allow media-associated contaminants to grow and cause a false positive result. If the temperature of the autoclave is too hot, overheating may have a detrimental effect on the media, denaturing it or charring necessary nutrients. This could create a situation where it would be difficult to detect stressed organisms. The FDA has observed that in many recalls for sterility, the investigation led to an initial sterility test failure.

The USP sterility test procedure calls for the testing environment to be equivalent to the aseptic processing environment. Proper design of the testing area thus includes a gowning room and pass-through air locks. Environmental monitoring and gowning should be equivalent.

The actual testing procedures should be documented as to their source, such as the USP and other microbiological references. Conformance to the procedure should be evaluated. Alternative test procedures to USP compendial methods may be used, but the methods must be validated properly.

The management of the microbiology laboratory will be a subject of the inspection. The contention is that the evaluation and interpretation of data requires extensive training and experience in microbiology. Understanding the technology and the limitations of the test becomes very important. If microbiological testing is done outside the company, the management's policy and performance on auditing the quality of the work of the contracting laboratory should also be part of the inspection.

By reviewing the inspection guides for FDA inspectors and doing a self-evaluation based on these guides, a company's QC/QA group can be better prepared for that inevitable inspection.

FUTURE TRENDS

The future role of QA and QC in the area of biotechnology will be one of increased involvement. The trend of emphasis will be more on the proactive, prevention mode. This increased tendency is demonstrated by the proposed amendment to the cGMPs. Proposed changes to the cGMPs were issued in the *Federal Register* on May 3, 1996, as an "Amendment of Certain Requirements for Finished Pharmaceuticals: Proposed Rule" (FR 1996e).

The changes that are proposed particularly emphasize the role of QA in validation. Several new definitions are included in the changes. Some of the changes have appeared in previous documents, such as guidelines and inspection guides, but they are now proposed to be an official part of the cGMPs. The proposed rule includes definitions for validation protocol, process validation, methods validation, equipment suitability, process suitability, out of specification, reprocessing, and manufacturing process.

- *Validation protocol:* "A written plan describing the process to be validated, including production equipment, and how the validation will be conducted, including objective test parameters, product and/or process characteristics, predetermined specifications, and factors which will determine acceptable results."

- *Process validation:* "Establishing, through documented evidence, a high degree of assurance that a specific process will consistently produce a product that meets its predetermined specifications and quality characteristics."

- *Methods validation:* "Establishing, through documented evidence, a high degree of assurance that an analytical method will consistently yield results that accurately reflect the quality characteristics of the product tested."

- *Equipment suitability:* "The established capacity of process equipment and ancillary systems to operate consistently within established limits and tolerances."

- *Reprocessing:* "A system of reworking batches that do not conform to standards or specifications . . . "

- *Manufacturing process:* "All manufacturing and storage steps in the creation of the finished product from the weighing of components through the storing, packaging, and labeling of the finished product, including, but not limited to, the following: Mixing, granulating, milling, molding, formulating, lyophilizing, tableting, encapsulating, coating, sterilizing, and filling."

- *Process suitability:* "The established capacity of the manufacturing process to produce effective and reproducible results consistently."

- OOS results were discussed in-depth in *the Guide to Inspection of Pharmaceutical Quality Control Laboratories.* They are an examination, measurement, or test result that does not comply with preestablished criteria as required by 21 CFR 211.160(b).

Increased Responsibilities for Quality Control

QC responsibilities are expanded in the proposed rule. In 21 CFR 211.22(a), the rule adds the responsibility for the quality control unit to review and approve validation protocols and to review changes in product process, equipment, or other changes to determine if and when revalidation is warranted. In many cases, this was a responsibility that companies assigned to the QA group, but now it will be a regulatory requirement.

The new 21 CFR 211.160, General Requirements, states that laboratory control should include

the establishment of scientifically sound and applicable written specifications, standards, sampling plans, and test procedures, including resampling, retesting, and data interpretation procedures designed to ensure that components conform to applicable standards of identity, strength, quality, and purity.

The responsibility for additional stability program oversight is given in 21 CFR 211.166. After the expiration date has been determined, an ongoing testing program for each drug product should be established to ensure product stability. At least one batch of each drug product should be added to the stability program annually.

Quality Assurance Responsibilities

QA responsibilities have been delineated for the review of production, control and laboratory records in 21 CFR 211.192. There will be a requirement for SOPs describing the QC review and approval of all drug product production, control, and laboratory records, including packaging and labeling, to determine compliance with all written procedures and specifications. SOPs for the investigation of any unexplained discrepancy or the failure of a batch or any of its components or in-process materials must be established. The investigations should include topics such as the following:

- Procedures to identify cause of the failure.

- Criteria for determining if OOS results came from a sampling or lab error.

- Sound procedures for excluding any test data.

- Procedures for additional sampling.

- Extending investigation to other batches.

- Review and evaluation of the investigation.

- Criteria for final approval or rejection.

- Written investigation reports.

- Reason for the investigation.

- Description of the investigation.

- Results of the investigation.

- Sound scientific justification for excluding OOS results.

- Conclusions and subsequent actions.

- Signatures and dates of person(s) responsible for approving the record.

- Signature and date of person(s) responsible for final disposition of the batch.

Validation Requirements

In the past, there were very few actual GMP references to validation. A new section is being added to specifically cover validation. This new section is Subpart L. Many of the concepts have been covered in other FDA documents, such as the process validation guideline or other inspection guides. QA has been given the responsibility for reviewing and approving validation protocols and reviewing changes in product process, equipment, or other changes to determine if and when revalidation is warranted. Therefore, this new subpart has a significant QA impact.

Subpart L addresses process validation and methods validation. The manufacturer will be required to validate all drug product manufacturing processes. These processes include computerized systems that monitor and/or control the manufacturing process and all manufacturing steps. Manufacturing steps include, but are not limited to, cleaning, weighing, measuring, mixing, blending, compressing, filling, packaging, and labeling.

Validation protocols must be prepared and followed that identify the product and product specifications and specify the procedures and acceptance criteria for tests to be conducted. The data that are to be collected during process validation will have to be developed and approved as part of the protocol. Protocols must specify a sufficient number of replicates to demonstrate reproducibility of the process and to provide a measure of variability between runs. The rationale for the number of replicates should be scientifically sound and justifiable. Validation documentation should also include standards for the suitability of materials used in the process and standards for the performance and reliability of equipment and systems involved in the processes. Through a mechanism such as validation reports, the manufacturer must be able to document execution of the protocol and test results. Both equipment and processes must be selected to ensure that product specifications are reliably met. The manufacturer must determine suitability, including tests to verify that equipment operates satisfactorily within operating limits, in meaningful qualification studies.

There must be a QA system that requires revalidation whenever there are changes in parameters such as packaging, component characteristics, formulation equipment, processes, *and* when changes are observed in product characteristics. QA will play a major part in monitoring the product characteristics.

In addition, there must be a program for methods validation in which the accuracy, sensitivity, specificity, and reproducibility of test methods are validated and documented. The requirement for validation protocols reviewed and approved by QA is the same as for manufacturing processes and equipment. Previously, the cGMPs required that methods be "established." The FDA has shifted and clarified the emphasis by changing the term to "validated." The intent of methods validation is to demonstrate that a method is scientifically sound and that it serves its intended purpose. In order to produce results that are reliable, meaningful, and trustworthy, the methods used to obtain the results must be reliable. In addition, the methods used to analyze the data obtained by these methods must be validated.

Contamination Control

Another new section is Subpart M—Contamination. Procedures for the prevention of contamination, including cross-contamination, will have to be established. This includes controlling the environmental conditions—facilities, air handling equipment, and/or process equipment. This implies the implementation of environmental monitoring programs and facility and air handling equipment validation. Contamination control can be achieved through a combination of proper design, cleaning processes, employee training, gowning procedures, and air filtration.

The FDA believes that by placing the responsibility for the oversight of validation on the QC unit, there will be a greater emphasis on the importance of validation to QC.

Biotechnology Product Quality

Another important topic in biotechnology product quality is the government definition of well-characterized biotechnology drugs as part of a series of regulatory reforms aimed at reinventing government (REGO) (Little 1996). One of the significant changes as a result of the REGO initiatives include the allowance of filing Establishment License Applications (ELAs) and Product License Applications (PLAs) at different times. The elimination of ELA and lot-to-lot testing requirements for well-characterized biotech products are important REGO changes. As a result, CBER will no longer approve establishments for the manufacture of a specific drug but will inspect to determine compliance with cGMPs. The newly revised PLA will require detailed chemistry, manufacturing, and control (CMC) sections instead.

INTERNATIONAL IMPACTS

The trend in quality is global. In order for companies to compete, they must be international in scope. The FDA supports the concept and, in fact, participates in international harmonization activities. The International Conference on Harmonisation (ICH) has been established to promote consistency in international regulations. ICH working groups have the task of harmonizing the various regulations from the United States, Europe, and Japan. Three technical areas on which the ICH groups focus are efficacy, safety, and quality. The quality area is comprised of five major elements: stability, validation of analytical methods, impurities, pharmacopeias, and biotechnology products. Each group is given the responsibility for drafting regulatory documents within its subject area. The documents go through a series of five steps before they are actually enforced. The first step is drafting the document, which is then reviewed, revised as necessary, and approved by the working groups. The approved document is then published in the *Federal Register* to solicit public comments and, after addressing public comments and issues, is subsequently approved by the FDA. The next step is approval by all participating countries. This results in the final document. It is a good plan to obtain copies of the regulations prior to their approval by the FDA, even though the regulations are not enforced by the FDA until this stage. The documents can be used to develop plans for the ultimate course of action that the company will follow once the document is implemented. The pharmacopeia group is responsible for the standardization of the Japanese, European, and U.S. compendial methods.

Some of the ICH quality documents that are currently active include two in the stability area: *Stability Testing of New Drugs and Products* and *Photostability Testing* (FR 1994, 1996c). The group addressing the validation of analytical methods also has two active documents that have appeared at different stages: *Validation of Analytical Methods: Definitions and Terminology* and *Validation of Analytical Procedures: Methodology* (FR 1995a, 1995b). The impurities group has three active documents in preparation: *Impurities in New Drug*

Substances, Impurities in Dosage Forms: Addendum to the Guideline on Impurities in New Drug Substances (FR 1996a, 1996d), and *Impurities: Residual Solvents* (in an early stage). The biotechnological products group is addressing products from cell cultures in its documents on viral testing (ICH 1995) and genetic stability (FR 1995c, 1996b). The topics of other documents recently approved for preparation include the stability of and specifications for test methods.

As noted previously, there will be increased emphasis on analytical methods—their development and their validation. Laboratory practices will be scrutinized more than ever.

REFERENCES

Code of Federal Regulations. 1995. Title 21, Parts 211, 610, and 820. Washington, D.C.: U.S. Government Printing Office.

DeSain, C. V. 1992. *Documentation basics that support good manufacturing practices.* Eugene, Ore., USA: Aster Publishing.

EEC Ad Hoc Working Party on Biotechnology/Pharmacy. 1991. Validation of virus removal and inactivation procedures. *Biologicals* 19:247.

FDA. 1985. *Points to consider in production and testing of new drugs and biologicals produced by recombinant DNA technology.* Bethesda, Md., USA: Food and Drug Administration, Center for Biologics Evaluation and Research, Office of Biologics Research and Review.

FDA. 1987a. *Guideline on the general principles of process validation.* Rockville, Md., USA: Food and Drug Administration: Center for Drug Evaluation and Research, Center for Biological Evaluation and Research, and Center for Devices and Radiological Health.

FDA. 1987b. *Guidelines on validation of the limulus amebocyte lysate test as an end product endotoxin test for human and animal parenteral drugs, biologics, and medical devices.* Rockville, Md., USA: Food and Drug Administration: Center for Drug Evaluation and Research, Center for Biological Evaluation and Research, and Center for Devices and Radiological Health.

FDA. 1987c. *Points to consider in the characterization of cell lines used to produce biological products.* Bethesda, Md., USA: Food and Drug Administration, Center for Biologics Evaluation and Research, Office of Biologics Research and Review.

FDA. 1993a. *Guide to inspections of pharmaceutical quality control laboratories.* Rockville, Md., USA: Food and Drug Administration, Office of Regulatory Affairs, Office of Regional Operations, Division of Field Investigations.

FDA. 1993b. *Guide to inspections of microbiological pharmaceutical quality control laboratories.* Rockville, Md., USA: Food and Drug Administration, Office of Regulatory Affairs, Office of Regional Operations, Division of Field Investigations.

FDA. 1993c. *Guide to inspections of validation of cleaning processes.* Rockville, Md., USA: Food and Drug Administration, Office of Regulatory Affairs, Office of Regional Operations, Division of Field Investigations.

FDA. 1994. *Points to consider in the manufacture and testing of monoclonal antibody products for human use.* Bethesda, Md., USA: Food and Drug Administration, Center for Biologics Evaluation and Research, Office of Biologics Research and Review.

FR. 1994. Stability testing new drug substances and products. *Federal Register* 59 (183):48754–48759.

FR. 1995a. Draft guideline on validation of analytical procedures: Methodology. *Federal Register* 61 (46):9315–9319.

FR. 1995b. Guideline on validation of analytical methods: Definitions and terminology. *Federal Register* 60 (40):11260–1262.

FR. 1995c. Draft guideline on stability testing of biotechnological/ biological products. *Federal Register* 60 (161):43501–43505.

FR. 1996a. Impurities in new drug substances. *Federal Register* 61 (3):371–376.

FR. 1996b. Final guideline on analysis of the expression construct in cells used for the production of r-DNA derived protein products. *Federal Register* 61 (37):7006–7008.

FR. 1996c. Draft guideline for photostability testing of new drug substances and products. *Federal Register* 61 (46):9309–9313.

FR. 1996d. Draft guideline on impurities in new drug products. *Federal Register* 61 (54):11267–11272.

FR. 1996e. 21 CFR Parts 210 and 211: Current good manufacturing practice: Amendment of certain requirements for finished pharmaceuticals: Proposed rule. *Federal Register* 61 (87): 20103–20115.

ICH. 1995. Viral safety document: Step 2. *Viral safety evaluation of biotechnology products derived from cell lines of human or animal origin.* Geneva: International Conference on Harmonisation.

Joner, E., and G. D. Christiansen. 1988. Hybridoma technology products: Required virus testing. *BioPharm* 1 (8):50–53.

Little, L. E. 1996. Focus on QC: Keeping on track: Current quality control trends. *BioPharm* 9 (5):72–77.

Rhodes, W. E. 1985. Monoclonal antibodies for in vivo use: In-process and final testing. *Pharm. Manuf.* 2 (4):12–14.

Schiff, L. J., W. A. Moore, J. Brown, and M. H. Wisher. 1992. Lot release: Final product safety testing for biologics. *BioPharm* 5 (5): 36–39.

USP 23/NF 18. 1995. *United States Pharmacopeia,* 23rd ed./*National Formulary,* 18th ed. Rockville, Md., USA: United States Pharmacopeial Convention, Inc.

4

BIOTECHNOLOGY MANUFACTURING ISSUES: A FIELD INVESTIGATOR'S PERSPECTIVE

Gregory Bobrowicz

Quintiles BRI, Inc.
(Field Investigator for the FDA
at the time this chapter was written)

The U.S. Food and Drug Administration (FDA) is the public health agency charged with protecting consumers from unsafe or ineffective drugs, blood products, medical devices, food, and cosmetics. The FDA is also responsible for the expeditious approval of new drugs and medical devices. To accomplish these missions, the FDA inspects the facilities where these commodities are manufactured to evaluate their compliance with Good Manufacturing Practice (GMP) regulations (CFR 1994). Generally, these inspections are conducted by field investigators. This chapter is written from the perspective of a field investigator.

Most of the biotechnology industry has not been exposed to the FDA field organization (the Field), since relatively few products have progressed beyond the research and development stage. Accordingly, the purpose of this chapter is to describe the general approach

---

*This chapter represents the opinion of one investigator and, as such, neither binds nor confers rights to the FDA or any other party.

during an inspection, emphasizing a field investigator's concerns. The Center for Biologics Evaluation and Research (CBER) has a formal, detailed managed review process (Paulson 1994; Wechsler 1994) for product and facility licensure. Licensed biologics, therefore, are subject to regulatory policies and procedures that are distinct from those in this chapter.

This chapter begins with a description of the Field's involvement in inspections of the emerging biotechnology industry. Generally, the inspectional approach is the same for biopharmaceuticals and traditional chemical drugs. The next sections discuss ways in which a product can become unsafe or ineffective. The final section addresses managerial tools to improve compliance. The principal focus of this chapter is to explain the FDA's reasoning and priorities, not to describe what the FDA may emphasize on a particular inspection.

REGULATION OF BIOTECHNOLOGY PRODUCTS

The Centers and the Field

For purposes of this chapter, the FDA consists of Centers and the Field (FDA 1994). The Centers regulate specific product lines: food (Center for Food Safety and Applied Nutrition, CFSAN), drugs (Center for Drug Evaluation and Research, CDER), veterinary medicine (Center for Veterinary Medicine, CVM), medical devices (Center for Devices and Radiological health, CDRH), and biologics (CBER). The Centers are located in the metropolitan area around the nation's capital, together employing about 4,300 people. They are responsible for the scientific and medical review of applications, and their compliance offices review regulatory cases for technical and legal sufficiency.

CBER regulates most biopharmaceuticals. It is ultimately responsible for product and facility approval, including medical, toxicological, and manufacturing process review. Unique among the FDA's headquarters organizations, CBER has historically conducted inspections of regulated firms. However, CBER will increasingly turn over inspections of biopharmaceutical plants to the Field organization.

The Field consists of about 3,000 employees distributed among a headquarters unit, 21 District Offices, and 133 geographically dispersed smaller offices. Field investigators conduct inspections of all regulated commodities worldwide. For drugs, biologics, and medical devices, field investigators inspect firms for compliance with GMP regulations. Field inspections of biopharmaceutical manufacturers have mostly been for biopharmaceuticals that are not licensed biologics (insulin and growth hormone, for example), clinical supplies for export, and commercial products inspected in conjunction with CBER. Accordingly, the perspective of this chapter draws heavily on these segments of the industry.

Problems Noted at Biopharmaceutical Plants

During inspections of biopharmaceutical firms, field investigators have noted problems in basic quality areas. The following list of common problems draws from the experience of multiple field investigators at various firms. The issues are presented in no particular order.

- Sterile filtration not validated.
- Environmental monitoring:
  - Personnel not monitored.
  - Data not reviewed.
  - Investigations superficial/not conducted.
  - Viables not monitored.
  - No identification of contaminants.
  - Alert limits set too high.
- Component testing:
  - Components passed on certificate of analysis and identity test only.
  - Certificate of analysis accepted without verification.
- Training:
  - Investigations do not lead to retraining.
  - Training not documented.
  - No training in particular operations.
- Materials not segregated:
  - Components.
  - Rejects.
  - In-process materials.
- Equipment calibration & maintenance:
  - Thermometers, pipettes, spectrophotometers, and autoclaves lack scheduled maintenance.

- Time limits not set:
    - Cleaning.
    - Restarting procedures after idle time.
    - Storage of intermediates.
    - Component expirations.
- Computer systems not validated:
    - Data management (environmental and samples).
    - Process control.
    - Lab operations (plate readers, spectrophotometers).
- Documentation (batch records and lab records):
    - Steps not signed.
    - Changes not explained.
- Review:
    - Superficial.
    - Not timely.

It is essential that this list be viewed as an instructional tool, not as a checklist. Each inspection is different, and each firm has unique problems. Nevertheless, it is clear from this list that there are problems with basic GMP compliance in the segments of the industry that the Field inspects.

One explanation for the fundamental nature of the problems noted by field investigators may be a misunderstanding on the applicability of the GMP regulations. Some firms erroneously believe that GMPs do not apply to production of clinical study materials. In addition to an ethical obligation to their patients, manufacturers of clinical study materials are required by regulation to comply with GMPs (CFR 1994; FR 1978; FDA 1991a). GMP compliance is also a good business practice, as the automotive, computer, and aerospace industries have found.

Another explanation for the fundamental nature of the problems noted in the biotechnology industry may stem from uncertainty in applying the GMP regulations to novel biotechnology processes. The drug GMP regulations cover the gamut of pharmaceuticals and biopharmaceuticals. In deference to the technical sophistication, speed of innovation, and variety of processes in this industry, the FDA has avoided a "cookbook" approach in the GMP regulations. Instead, it has promulgated broad regulations requiring application under sound scientific principles. To assist manufacturers in complying with GMP regulations, the remainder of this chapter will discuss the FDA's rationale for inspectional emphases and suggestions for comprehensive, rather than incremental, approaches to compliance.

FDA's CONCERNS

A comprehensive quality guide to every aspect of biopharmaceutical manufacturing is beyond the scope of this chapter. This chapter also assumes that the reader has a sound understanding of quality control (QC) principles. Indeed, since field investigators have noted primarily fundamental problems at biopharmaceutical companies, it would be pedantic to recite them. It would also be tedious to explain each functional or regulatory area in the plant in light of the Field's expectations. Instead, this chapter explains that the FDA bases its inspectional emphases on scientific evaluation of potential health hazards, not rote adherence to policies and regulations.

While flexible, the FDA's expectations are not arbitrary; scientific reasoning and expertise are foundations for its regulatory positions. Because the FDA is primarily a public health agency, it designs its requirements to protect consumers from unwarranted risk. Rather than covering each area of a firm, Field investigators target their inspections on critical control points that reasonably connect to patient health and welfare. Secondarily, the FDA is responsible for assuring a level playing field and protecting the consumer's pocketbook. Potential public health problems in biopharmaceutical manufacture may include impurities related to the drug substance, chemical impurities, adventitious agents, pyrogens, potency variations, and misidentification of the product. Critical control points for each of these potential hazards are discussed below.

Related Impurities

Related impurities are so named because they are related to the target product (USP 1995, pp. 1922–1924). Related impurities in biopharmaceuticals have a great potential for pharmacological activity. Even inactive impurities, such as denatured proteins, may be problematic, since it may be difficult to distinguish them from the active substance, a confusion that may result in a subpotent dose.

There are sources of intrinsic heterogeneity in proteins from imperfect transcription, translation, and glycosylation. Generally, the levels of impurities from biological processes are below the level of concern. The more significant sources of impurities result from problems in fermentation and purification. Field investigators' concerns about impurities increases with proximity to the finished product. The remainder of this section will discuss inspections of downstream processing.

Purification

It is important to characterize the feed stream to any purification process. The purification process is designed to remove known impurities; major changes to the fermentation process, including contamination with microorganisms, may introduce unknown contaminants for which removal has not been demonstrated. Validation of fermentation processes has been discussed elsewhere (Nelson and Geyer 1991; Diers et al. 1991; Muth 1991) and will not be discussed here.

The initial removal of the majority of impurities is relatively straightforward. However, getting the last 5 percent of impurities down to 0.5 percent is more challenging. To date, chromatography has been the only commercially viable technology for the purification of proteins. Investigators examine chromatographic purification closely because of the importance and complexity of these steps.

The chromatographic medium should be well characterized and tested. Early in the process, a manufacturer is expected to determine and control the critical attributes of the medium. For some manufacturers, these attributes have included particle size distribution, extractables, and microbiological load (Seely et al. 1994; Barry and Chojnacki 1994). Acceptance specifications can duplicate the manufacturer's certificate of analysis, but the customer should verify the certificate independently. By the time a process approaches large-scale clinical trials, the chromatographic medium should be fully characterized. For characterization and acceptance testing, there are no tests superior to functional testing under lab-scale conditions that mimic the process.

Functional testing of the purification column is essential, but it is insufficient to rely solely on functional testing to ensure that the column is free of defects (Barry and Chojnacki 1994). A manufacturer should have detailed procedures for setting up the production column. These procedures should include all of the parameters necessary to establish that a column is performing properly. The FDA's major concern is that inadequately characterized chromatography will result in unexpected and, hence, uncontrolled impurities.

Process-Related Modifications

Proteins are complex organic molecules with a variety of possible degradants. Depending on such factors as temperature and pH, proteins may undergo β-elimination, oxidation, deamidation, or proteolysis, among other modifications (Klegerman and Groves 1992, p. 208). A manufacturer should identify and control the critical parameters causing degradation. Some firms elect to compile these studies in a development report.

Metal Ions. The detrimental effects of metal ions on proteins are well documented (Stadtman 1990). The sources of metals should be identified and controlled prior to production. All product-contact surfaces, including column parts and storage tanks, should be evaluated for reactivity prior to purchase and installation. Unless a system is very new, an investigator may make inquiries into the design specifications and construction of the process train, concentrating on the possibility of corrosion and wear. Particularly when equipment is exposed to corrosive solvents and detergents during cleaning, the quality of the metal finish may need to be monitored regularly.

When proteins are known to be sensitive to metals, components should be tested for metal content. A biotechnology manufacturer may need to supplement the standards in the compendia with specifications for metal content or other impurities.

Agitation and Heating. Proteins may be denatured during exposure to heat, agitation, and phase interfaces, such as with air (Sadana 1989). To minimize denatured proteins in the product, it is important to control the flow of product solution throughout the process. During a facility inspection, investigators may determine if there are liquid/air interfaces and what the manufacturer has done to prevent denaturation at that interface. In early clinical stages, when the balance of quality assurance (QA) comes from equipment qualification and extensive product testing rather than validation, a firm should have temperature mapping data to ensure that there are no local "hot spots." Process steps with high agitation, such as pumps and fillers, must be shown to have minimal adverse effects. A manufacturer may test a sample of the product after passing it through a "worst-case" agitation; such a sample may be obtained from the end of a filling operation after maximum agitation.

Regardless of the stage of process development, documentation of purification processes is essential to ensure quality. Changes in equipment configuration and operation should be justified. Once the process is scaled up to commercial batch sizes, investigators will compare the actual manufacturing conditions to the previously validated ranges.

Testing

Regardless of how related impurities may accumulate in a process, a manufacturer needs sensitive analytical methods to detect them. Technology for these sensitive assays is likely to continue to improve, and the FDA's expectations are likely to respond to industry standards. Accordingly, it is difficult to discuss pitfalls of particular methods, since the problems, and possibly the methods themselves, will soon be obsolete. Laboratory methods required to distinguish impurities are described elsewhere (USP 1995, pp. 1849–1859; Klegerman and Groves 1992, pp. 175–202; Garnick et al. 1991). However, FDA has some broad expectations, as described below.

The language of the method should be clear to the QC analysts responsible for performing the method. In many cases, research labs write vague or uninterpretable instructions. For example, one chromatographic method included a "recommended system suitability" that the laboratory analysts felt free to disregard since it was only a recommendation. The inspection team was not as charitable. Methods performed outside the parameters generated during methods validation are unreliable.

During methods development, the laboratory should consider whether the method can be transferred to a commercial lab. In some methods, solvents incompatible with certain high performance liquid chromatography (HPLC) columns have destroyed the columns, which is unacceptable in both a research setting and a regulatory environment. Similarly, equipment must be commercially available; readily available equipment is preferred. If equipment is customized, the firm risks the possibility that the equipment will be unavailable at some future date. In the same vein, some viral methods and impurity tests are simply too delicate to transfer outside the research lab. A firm should carefully consider designing methods during the earliest stages of product development.

In reviewing methods validation data, the FDA frequently finds that laboratories have failed to assess the following:

- Sample interferences from excipients (placebo or stressed placebo).

- Method conditions (injector heat or filters that interfere).

- Deterioration of the sample on standing in solution.

- Potentially unresolved degradants.

- Interferences in ultraviolet (UV) spectra (a single point is insufficient).

Reviewers in the various Centers will evaluate methods validation as well. For biologics, the division of responsibility between CBER and the Field has not yet been formally delineated for this task.* Traditionally, however, the Field organization has greater access to raw data, and the Centers have broader technical insights.

Laboratories should characterize reference standards based on purity. In addition, most methods require that standard concentrations should approximate the level expected in the test article. Careful documentation of the reference article is critical during methods development and validation.

Like most GMP problems at biopharmaceutical plants, problems in the development of laboratory methods are often basic oversights. Data supporting an official method must be generated in accordance with GMPs; in some cases, data are not adequately documented and reviewed. One firm's chemists turned off the HPLC chart recorder before the emergence of a potential degradant; the supervisor misleadingly reported the result as "none detected."

It is important to characterize fully a compound being tested in human subjects. A firm must establish that the molecular species has not changed since early clinical trials. Accordingly, critical methods, such as assay and characterization of key impurities, should be developed and validated before the start of clinical trials to allow accurate interpretation of clinical data. Sensitive and reliable methods are the foundation for critical manufacturing decisions as well.

Stability

Stability testing of a biopharmaceutical must include testing of both chemical and physical parameters (Manning et al. 1989). Stability programs should also address reconstitution of lyophilized proteins. The FDA has provided detailed guidance in conducting stability studies (FDA 1987b), and the International Conference on Harmonisation (ICH) has addressed stability as well (USP 1995, pp. 1959–1963).

Investigators frequently have found problems in stability studies. In some firms, the raw data are not meaningfully reviewed, and final results are not trended. In other firms, the stability sampling

---

*The FDA's NDA/ANDA Preapproval Inspection Program describes the division of review responsibilities for human drugs; this program does not apply to biologics.

programs have failed to represent a production lot statistically. Investigators have criticized the practice of averaging failing results into compliance. However, the greatest problem is in methods development. Stability methods must be shown to be "stability indicating," that is, degradants must be identified and quantified. Investigators will carefully review stability data to ensure that stability reports are complete and that the firm followed a standardized protocol.

Viruses

Viruses raise a risk of an "Andromeda strain": an unknown potent pathogen. The concern is not merely theoretical; there are historical examples of contaminated biologicals leading to gravely nonfictional tragedies (Grun et al. 1992). Biopharmaceuticals potentially expose humans to viruses that ordinarily do not mix with human blood. This risk is the motivation for the extreme conservatism of viral elimination. Furthermore, it is difficult to test for an unidentified, extremely rare hypothetical contaminant. Therefore, removal and inactivation steps should be general rather than specific to a single virus type. The strategy for viral reduction is threefold: eliminate sources of viruses; test for viruses in sources, intermediates, and product; and validate the removal and inactivation of viruses at high titers.

Sources

Viruses may enter the process stream from the cell line, components, the environment, or equipment. Of these, the greatest concern is with the cell line and serum. The cell line should be screened in accordance with FDA guidance (FDA 1993f). Serum must be tested; like any component test, it is insufficient to rely solely on a vendor's certificate of analysis in the absence of independent verification.

Testing

Testing programs are generally decided in concert with the CBER's review division, and the Field will generally not be concerned with the commitments in the application. At a minimum, most firms thoroughly test the master cell bank (MCB) and working cell bank (WCB), the cells at the end of fermentation, and the finished product. All of these tests must be conducted in compliance with GMPs, including maintaining associated records and validation documents. Test methods must be fully validated; many biotech products modify cellular actions to infection, so it is essential to demonstrate that a given product does not interfere with an assay's ability to detect a positive result.

Clearance

Most removal and inactivation processes, particularly chromatography, are highly variable among both viruses and processes (Grun et al. 1992). The selection of model viruses should be based on health risk, variety, and actual likelihood of occurrence. A process should be able to remove extremely pathogenic viruses, even if their presence is unlikely. Some viruses, such as human immunodeficiency virus (HIV), are too dangerous to handle in high titers; it may suffice to bracket the pathogen with model viruses. Validation studies should include many different model viruses of varied sizes, genetic material, and encapsulation. Validation of virus removal and inactivation studies should be done early in process development. It is desirable to include a single process step that generically reduces the concentration of all viruses by at least five log units (White et al. 1991).

Manufacturers should consider the kinetics of viral inactivation to justify linear extrapolation from high titers. Scaled-down processes should be shown to be substantially equivalent to the full-scale process. FDA investigators may verify test methods and results against filed commitments.

Mycoplasma

Control of mycoplasma is similar to viral control in several respects. Principally, the nature of the risk is similar. First, mycoplasma are quite able to pass through conventional 0.2 μm sterilizing filters. Second, mycoplasma contamination may come from eukaryotic cell culture or animal serum. Third, both contaminants are potentially deadly. The general characteristics of mycoplasma, including detection and removal, are reviewed elsewhere (Roche and Levy 1992).

The strategies to minimize mycoplasma contamination derive from their general characteristics. As described in an FDA publication (1993f), there should be tests for mycoplasma in the MCB, WCB, and possibly also the concentrated harvest pool. A manufacturer should also include mycoplasma testing in the lot release specifications for serum. As with all component tests, it is insufficient to rely solely on a vendor's certificate of analysis. Field investigators may audit raw data for these tests to determine if a firm is meeting its filed commitments.

Microorganisms/Endotoxins

Product quality requires minimizing microbial contamination of the process stream. First, microbial contamination of an injectable drug product can be fatal. Second, microbes can degrade proteins, excipients, and chromatographic columns. Third, uncontrolled bacterial growth may result in elevated endotoxins, causing adverse reactions in patients. Accordingly, Field investigators concentrate heavily on microbial quality. The Field's concern with product contamination increases throughout the process; the greatest concern is with the finished product.

Microbial quality depends on the specific process and general systems of the plant. All facility systems (e.g., water; heating, ventilation, and air-conditioning [HVAC]; sterilization systems; and training) need to be qualified when the product is first used in the clinic. In contrast, the specific manufacturing process is refined until it is finally validated at product approval. The validation of systems and specific processes is discussed separately below. The organization of this section is intended to be philosophical rather than technical.

Process Related

Manufacturers make great efforts to minimize endotoxin levels. Nevertheless, endotoxins are common in biological processes, particularly in prokaryotic systems. A manufacturer must validate the removal of endotoxins. Some manufacturers have unsuccessfully attempted to demonstrate a priori that a process intrinsically minimizes endotoxins. Sufficient validation of endotoxin removal includes worst-case testing. In-process testing for endotoxins and bioburden ensures that endotoxins do not exceed the levels removed during validation studies. Excipients in the final formulation, containers, closures, and key intermediates should be monitored for endotoxins with an appropriately validated method (FDA 1987c; USP 1995, pp. 1696–1697).

Similarly, a manufacturer must have data validating the sterilization process of each specific product. When filtration is used, filter test conditions must be linked to microbial retention; some firms elect to contract with a filter supplier to conduct the tests that demonstrate effective retention in a specific product. In any event, it is critical to show that particular parameters (flow rate, pressure, temperature) and filter attributes (bubble point, pressure hold) are sufficient to remove microbes from the product solution (FDA 1987a). Some filter manufacturers have argued that a sterilization process can be validated based on sterilization data from similar processes and proteins; the Centers, not the Field, will make the final decision to accept or reject this approach for a given product.

Time Limits. Some manufacturers have left intermediates for weeks at room temperature without assessing the effect on the product;

this is objectionable because intermediates usually approximate physiological conditions that are ideal for microbial growth. Similarly, even sterilized equipment and components do not necessarily remain sterile indefinitely. To minimize contamination and growth, a manufacturer must establish resident time limits. The FDA recognizes that a manufacturer may not have enough experience to set time limits when manufacturing Phase I supplies. Relying on similar processes for other products is an acceptable alternative for these early stages (FDA 1991a). By the time a process is scaled up to produce commercial quantities, time limits should be set for all critical process steps.

In response to FDA comments, some manufacturers have maintained that time limits are set by limitations on the process. For example, a step that must be completed within a single shift is limited to eight hours, and column-loading that cannot be interrupted is bounded by flow rate and volume. While such responses are sometimes sufficient, manufacturers are encouraged to formalize these processes. Investigators have found instances where the batch records show operators interrupting a step despite the engineer's unwritten intentions. In the absence of specifications on time limits, a QA auditor reviewing the batch record does not find such an interruption to be a cause for concern.

Bioburden. Even after time limits are set and validated, bioburden should be monitored at key steps or after long storage. Contaminants should be identified at least to the genus level. Manufacturers rely on these identifications to understand the risk of any given contamination. Initial specifications should be derived from common sense and historical data. When bioburden is high, investigators will carefully evaluate the decision to continue manufacturing. Ultimately, concern will focus on potential endotoxins, unknown impurities, and modifications to the drug substance; the manufacturer should be prepared to provide scientific reasoning to explain the bioburden testing program with regard to these concerns. Although not required by regulation, a development report may address testing parameters, such as acceptance limits, test intervals, and sampling locations.

Sterilization. Sterilization is arguably the most critical step in the manufacture of an injectable biopharmaceutical. The FDA has written guidance for the industry (FDA 1987a) as well as guidance for its investigators (FDA 1993b). Indeed, there is no shortage of authoritative standard references on the sterilization of parenterals

(USP 1995, pp. 1963–1981; Phillips and O'Neill 1990; Avis and Akers 1986). Nevertheless, there are scores of recent recalls and countless adverse inspectional findings because of deficient sterilization assurance. The deficiencies in this area tend to fall into the categories described under "Strategies for Compliance" (detailed later).

Purification. Purification usually includes an endotoxin removal step and should be validated. Additionally, a manufacturer must limit the endotoxin and bioburden potentially contributed during purification. Microbial growth can occur on a column or during storage of intermediates.

To guarantee that purification steps do not contaminate the process stream, column sanitization should be validated to ensure low bioburden (Adner and Sofer 1994; Jungbauer et al. 1994). In addition, endotoxins and other contaminants must be removed from the column during cleaning; such cleaning may be especially challenging for the initial column, which may become clogged with cellular debris.

Column storage conditions should minimize growth. Storage conditions must be established, controlled, documented, and reviewed. The environment around the columns should be classified and controlled during operation and storage. Field investigators will review procedures and records for assembly, sanitization, testing, and storage of columns, emphasizing measures designed to prevent uncontrolled contamination.

Filling. Field investigators will spend considerable time reviewing filling operations because there are minimal controls between the fill and the patient. Filling operations expose the product to countless hard-to-clean surfaces, including the filling head, transfer lines, and the container-closure system. In addition, product is exposed, however briefly, to the air. Accordingly, environmental controls for the air system are critical. Statistically based action limits should be set for personnel monitoring, airborne viables and particulate matter, surfaces, and utensils.

FDA training courses have emphasized the value of watching the filling line for an extended period. If a filling line stops frequently, product is exposed to the air for a longer time. If the stoppages result in machine adjustments, then engineers or their tools may be in proximity to the open vials as well. Investigators will review employee practices, including gowning. Repeated handling of equipment (forceps, wash bottles, or machinery parts) is an activity of great concern. Investigators may review filling records looking for deviations from the process, unexplained rejects, prolonged stoppages, and environmental monitoring.

The FDA will closely scrutinize manual fill operations because of the risk of contamination from operators. Field investigators commonly criticize the environmental monitoring program on manual fill lines.

Lyophilization. Manufacturers should know that the FDA will scrutinize lyophilization operations closely. The FDA has published detailed guidance for the inspection of lyophilization of parenterals (FDA 1993c). This document emphasizes familiar themes for preventing microbiological contamination. Starting material must have a low bioburden. Equipment and facilities need to be qualified. This qualification includes environmental and personnel monitoring. In addition, a manufacturer should address common problem areas, including sterilizing the condenser and parts that move during the cycle, such as the internal stoppering mechanism. Facilities should be designed to optimize product flow under a controlled environment. Validation of sterilization should include media fills conducted under worst-case circumstances.

The inspection guide points out that the lyophilization process is demanding on equipment. A manufacturer should establish frequent tests to ensure proper performance of lyophilizers and include provisions for atypical circumstances, such as refrigerant leaks, air leaks, or power failures.

Systems Related

Facility. Inspections traditionally begin with a walk-through of the facility. During this initial overview, investigators are looking at employee practices, housekeeping, material flow, and the building itself. The plant layout affects the likelihood of contamination or mix-up.

Product and personnel flow should minimize the possibility of contamination. The design of the building should prevent personnel moving from dirty to clean areas without regowning. Similarly, materials should be moved with appropriate concern for microbiological contamination; for example, a dirty cart should not go into a Class 100 area. Investigators may review floor plans and process diagrams; investigators may ask for formal rationale for unusual flows of air, product, or personnel.

The airflow within the facility should move air from cleaner to dirtier areas. Airflow is also a critical factor in minimizing cross-contamination when containment is an issue. Airflow should be monitored continuously, with out-of-limit alarms set appropriately. Investigators will review procedures and historical data for calibration of air pressure meters, alarms, and responses to alarms.

Water Monitoring. Water is a good medium and source of microbial contamination. Accordingly, FDA inspections continue to emphasize water systems, and investigators continue to note problems. To be in control of a water system, a firm must have an as-built drawing detailing exactly how the system is configured. Investigators will compare the as-built drawing with maintenance records and personal inspection of the system. Investigators check for undocumented dead legs. When there are changes, the fittings and welds must conform to the original design specifications. In some cases, manufacturers have added drops to the water system with threaded fittings that are notoriously difficult to maintain. To maintain control of the water system, a manufacturer needs a formal change control procedure.

High-purity water systems should be monitored for chemical and microbiological contaminants. Investigators will ask for raw data as well as data summaries. While reviewing the data, an investigator may also check management's awareness and response to trends or actionable limits. Data from water systems are frequently filed without significant trending or review. This practice can result in undiagnosed chronic contamination.

The FDA has published specific and detailed guidance on inspections of water systems (FDA 1993a). The inspection strategy emphasizes

- evaluating data summaries,

- taking microbiological data in context,

- reviewing maintenance records, and

- confirming as-built diagrams.

Environmental Monitoring. Environmental monitoring data remain an inspectional focus. Often, firms neglect the environmental monitoring program; there is a failure to review the program to evaluate the need for investigations or improvements. Extensive environmental data set the baseline for the validated state. Excursions beyond baseline values demonstrate that the total system may not be in control.

Manufacturers should base action limits on common sense and statistical evidence. Sampling is unlikely to capture the highest level of contaminants in a facility because the samples are small and infrequent compared with the total airflow and surface area in a facility. Therefore, investigations should follow any unusual result. One inspected firm had no action limits on viable particles other than the room's specification; no investigations followed individual excursions, with counts exceeding the normal variation by as much as 18 standard deviations. In another instance, unusually high results were averaged over time, masking the true magnitude of the excursion. A third firm did not investigate excursions unless the results were repeated during the next run; such a system does little to ensure control during a single run.

Investigators will look at historical data for environmental monitoring in all critical areas. They will generally request both original laboratory data and summary data to evaluate management's awareness of the program. The testing schedule will be reviewed, including sampling times and locations, to ensure that the testing program will find excursions when and where they happen.

Equipment. Aseptic operations require highly reliable equipment. Investigators will review calibration and maintenance of all equipment (autoclaves, fillers, lyophilizers, integrity testers, clean-in-place [CIP]/steam-in-place [SIP] systems) to ensure that equipment has been functioning properly. Sometimes, nonroutine maintenance has gone undocumented, frustrating an auditor's attempt to evaluate the problem on lots made before the correction.

Temperature is important for filter tests. At one firm, filters were wetted directly from the hot side of the Water for Injection (WFI) system and tested without documenting a stable temperature. Another firm stored filters in the cold room and wetted them with room temperature water. Obtaining valid results for this critical test requires monitoring and recording the temperature of the test (Scheer et al. 1993). Ultimately, the pressure-based tests are surrogate markers for microbial retention; a manufacturer must have data correlating integrity tests with retention of *Brevundimonas (Pseudomonas) diminuta* or equivalent (USP 1995, pp. 1979–1981).

Aseptic operations require strict adherence to procedures with minimal human involvement. Most manufacturers are turning to automation in downstream processing. Automated systems are clearly superior to human systems for many tasks. However, computers lack intuition and discretion. A human will not attempt to add a negative quantity to a batch on February 31, for example. Therefore, software needs to be validated, with the level of validation appropriate for the task it is controlling. The FDA will audit the essential areas of software validation: establishing complete detailed specifications, enforcing established requirements for writing the code, testing the specifications systematically, documenting and reviewing the test data, and administering rigorous change control. Additional information is available in various government publications (FDA 1983).* The FDA will continue to emphasize quality control in computer systems.

Media Fills. Since the USP methods for sterility testing detect only gross contamination, media fills must demonstrate adequate process control. Aseptic processes must be validated at the earliest stages of process development. There is no waiver for clinical supplies. Both clinical and commercial manufacturers frequently have trouble demonstrating that media fills reflect the actual process.

In one case, a manufacturer of clinical supplies monitored personnel for microbiological contamination during media fills but not during product runs. In another firm, the media fills in a lyophilizer had far fewer units than a complete commercial run. In a third firm, the most highly trained shift conducted media fills, but the product runs were conducted over multiple shifts with less experienced individuals.

The FDA has provided specific guidance in the aseptic processing guide (FDA 1987a). For clinical runs of less than 3,000 units, the maximum number of units of media filled should be at least equivalent to the maximum batch size (FDA 1993c). Times, volumes, pressures, number of operators, and operator training should be the same as actual runs. Indeed, the validation conditions set during the media fills should set boundary values of the validated state for product fills, environmental microbial load, training, number of units, time of exposure, and so on.

Quality Assurance for Injectable Biopharmaceuticals

Visual Inspections. Manufacturers must perform a 100 percent check for particulate matter in injections. If this check is automated, the FDA may review the computer system's validation. Manual inspection operations will be reviewed closely. Many manufacturers formally validate the inspection process. However, FDA investigators

---

*The FDA's expectations in this area follow general quality requirements for the computer industry. These general requirements are described in NBS Special Publication 500-75: *Validation, Verification, and Testing of Computer Software* (U.S. Department of Commerce, National Bureau of Standards [1981]). Government standards are reviewed in *Good Computer Validation Practices: Common Sense Implementation* (Stokes, T., Branning, R. C., Chapman, K. G., Hambloch, H. J., and Trill, A. J. [1994], Interpharm Press, Buffalo Grove, Ill., USA).

have uncovered problems with the inspection process through field complaints of gross particulate matter and empty units.

As with any QC test, the results of a 100 percent check should be recorded. Any inspection should have a specification on the allowable number of rejects. Defects should be documented, classified, and reviewed. It is unacceptable to inspect and reject a significant portion of the batch without explanation.

Theoretical Yield and Accountability. The concept of calculating yield involves two subordinate calculations. *Accountability* is the sum of the process output, including rejects, samples, start-up waste, and acceptable product. Many quality firms also monitor the amount of acceptable material, which could be called *output.*

In one sterile operation reviewed, the allowable yield (accountability) was 90–105 percent, an unjustifiable range considering that the firm consistently had less than 0.5 percent deviation in accountability. Investigators note significant problems in *output.* One firm inspected out significant defects during a manual operation and recorded the reasons for the rejects on a special form intended for the area supervisor. Under this system, the number of units recorded in the batch record did not include in-process rejects. In another firm, accountability was high, but rejects totaling over 10 percent of a filling run were unexplained in the batch record; a reject rate of this magnitude for finished goods should have a documented explanation. A 10 percent loss of the lot because of low fill volume is far less significant than such a loss from particulate contamination. These concepts apply to any pharmaceutical manufacturing process, including purification, filling, and labeling.

Process Contaminants

Process contaminants, according to the USP, include reagents, raw materials, and solvents that may be introduced during manufacturing or handling procedures. The USP describes how limits are determined for clinical and commercially marketed materials (USP 1995, pp. 1922–1924). Process contaminants are distinct from related impurities because contaminants are not chemically similar to the target molecule. Field investigators will be concerned that

- there are limits on contaminants,

- test methods are appropriate for the potential contaminants, and

- procedures are adequate to prevent unexpected contamination.

Process contaminants result from inadequate purification, impure components, reactive closures, or inadequate cleaning.

Purification

A manufacturer should have development data (which may or may not be assembled into a product development report) describing the known potential sources of contaminants. Some examples are listed below, but it is important to understand that investigators review a firm's internal quality systems for identifying and proactively minimizing contamination for its specific process.

For example, the FDA is concerned about deoxyribonucleic acid (DNA) in a finished pharmaceutical (FDA 1993f). In consultation with reviewers in the Centers, a manufacturer should set limits on the allowable levels of DNA at critical stages. For eukaryotic cells, a limit will be set in consultation with the review division. For prokaryotic systems, process validation is an adequate replacement for final product testing for DNA. Investigators may review in-process and finished product test results for compliance with established standards. Investigators will review methods validation. The FDA expects that manufacturers will monitor other signal impurities in addition to DNA.

Processing steps can also introduce contaminants. For example, deficient cleaning can leave detergent on processing equipment (discussed below). A firm should evaluate chromatographic resins for leaching. Manufacturers should evaluate extractables from resins and other product-contact surfaces. Pipes, hoses, and tanks should be made of materials with minimal sorption and leaching; a manufacturer should specifically evaluate them to ensure compatibility with the product and the process during operational qualification.

Investigators may review equipment maintenance. In extreme cases, contaminants have come from lubricants and coatings on equipment. It is somewhat common to see contaminants from gaskets or shavings from abraded parts. Involvement of the quality unit, including GMP training, should extend to personnel performing maintenance and repairs.

Components

Many components in the biotechnology industry are unusual; some are unique to a single firm's process. Other components are well known in applications outside the pharmaceutical industry, where QC is less stringent. Thus, there are no compendial standards for many processing aids and excipients. Even the official compendia may not list all of the tests necessary to ensure that a component has all of the quality attributes necessary for a specific industrial process.* A manufacturer must evaluate all excipients for the potential risk they contribute to the user.

Container/closure systems are a common source of contamination (Wang and Chien 1984; Avis and Akers 1986). Rubber stoppers can release particles into a parenteral. Rubber and plastic stoppers have released salts or small molecules into products. In one case, a glass manufacturer deleted a finishing step to optimize its process. The resultant glass had a lower pH and caused stability failures (Schier 1995). In another case, a stopper released antioxidants during lyophilization, leading to particulate contamination of the finished dosage form. Containers and closures should be tested rigorously with the understanding that they are among the most likely sources of chemical contamination. Investigators will review these data, which may reside in a product development report, if one is prepared. Investigators may also evaluate the level of control that an applicant has over a vendor's manufacturing procedures for closures, noting whether the agreement has provisions for regular auditing and change control.

Cleaning

Cleaning is a major concern in the biopharmaceutical industry, and the FDA continues to find problems in this area. The FDA has provided guidance to investigators for general cleaning procedures (FDA 1993e) and for the biotech industry in particular (FDA 1991b). Cleaning is a critical process that should be validated early in product development. Failure to plan for cleaning validation has delayed approvals for many manufacturers, some of whom have needed to reengineer their manufacturing processes before they can successfully validate cleaning.

The most conservative approach to eliminate cross-contamination is to dedicate equipment to a particular active ingredient. This approach is unpopular because of the cost. However, when equipment is critical and difficult to clean, dedication may be the only option. Some FDA speakers have publicly recommended dedication of

---

*"While one of the primary objectives of the Pharmacopeia is to assure the user of official articles of their identity, strength, quality, and purity, it is manifestly impossible to include in each monograph a test for every impurity, contaminant, or adulterant that might be present . . . Tests suitable for detecting such occurrences, the presence of which is inconsistent with applicable manufacturing practice or good pharmaceutical practice, should be employed in addition to the tests provided in the additional monograph." (*The United States Pharmacopeia,* 23rd ed., p. 7)

filling heads, for example. A manufacturer should, at a minimum, plan campaigns to minimize the number of changeover procedures.

Since a contaminant is not necessarily uniformly distributed in the contaminated lot, product testing alone does not ensure a product's purity. A manufacturer must validate cleaning processes for all nondedicated product surfaces. In addition to minimizing microbiological contamination, cleaning validation consists of the following.

First, a manufacturer must establish pharmacological and toxicological limits on drug and detergent residues. For the most part, Field investigators will not evaluate the limits per se, but they may evaluate the means by which the limits were set. These limits should be derived from reasonable scientific rationale and pharmacological data. Laboratory studies should also show that detergents or solvents will remove product residue.

Data to support analytical methods and sampling techniques must ensure accurate recovery and quantification of these residues at the acceptance limits. These studies should address potential interferences of the swab and detergent.

A team of engineers, microbiologists, production employees, and QA representatives should standardize sampling locations for each piece of equipment. The team should consider the criticality of the location and difficulty of cleaning.

Investigators will review training records for operators in the specified cleaning procedures. Equipment should be soiled and cleaned under conditions of actual use. Independent QA employees should take samples at the worst-case locations according to established protocols. The results of these tests will determine the need to modify the cleaning procedure. The cleaning procedure is considered validated when a determined number of consecutive lots (often three) is successful. Regular reevaluations ensure that efficacy of the cleaning procedure has not drifted. If campaigns are long enough that training may lapse between product changeovers, it may be prudent to evaluate cleaning efficacy at the end of each campaign.

Time limits should be set and validated for the times that equipment may stand before cleaning and after sterilization.

The regulations do not specifically state whether a firm should use swabs or rinse samples. Similarly, the regulations do not specifically rule out a nonspecific method, such as total organic carbon (TOC). The FDA has historically preferred to see swabbing and test methods that are specific to the target chemical. However, if a manufacturer can provide data to prove that the recovery methods and analytical tests are at least as sensitive as the FDA's preferences, the FDA will consider alternative approaches.

Potency

Dosage

A dosage form must deliver an accurate amount of a drug when it is administered to a patient according to the instructions on the label. During filling, manufacturers should document statistically based fill volume/weight checks. Investigators will review the validation of filling to ensure uniform dosing; important parameters include fill rate and lot size. Investigators may review batch records to determine if a run required frequent fill-volume adjustments.

Inadequate lyophilization can result in a partially insoluble cake. Validation, lot clearance, and stability studies generally include a specification on dissolution to address this potentiality. Investigators will review these data, particularly validation data. The FDA's *Guide to Inspections of Lyophilization of Parenterals* (1993c) discusses factors affecting dissolution. Samples taken to assess dissolution and stability should be selected from throughout the lyophilizer.

Early clinical trials assess the toxicity and efficacy against a standard dose. Prior to approval, methods validation should be prioritized; the most important methods are those that ensure safety, followed by assay methods. Failure to validate assay methods adequately can delay approval by confounding interpretation of clinical studies and product development data. Data supporting release of clinical material or submissions to the FDA should be generated with validated methods; unvalidated methods produce unreliable results. Standard methods, such as pH or loss on drying, require minimal method development (USP 1995, p. 6). Compendial methods do not require method validation, but a manufacturer must show that the compendial method works with the specific test article (CFR 1994). Nontraditional assay methods, including enzyme-linked immunosorbent assay (ELISA) and radioimmunoassay (RIA), should be validated according to the guidelines in the USP (USP 1995, pp. 1982–1984). One manufacturer of clinical supplies applied an ELISA method for finished product assay with errors of 33 percent between replicates; this is objectionable.

Stability

Stability studies ensure a fully potent dose throughout the intended shelf life of an article. The stability of protein drugs depends on many of the same considerations as chemical drugs. However, proteins are also labile to proteolysis, denaturation, and chemical modifications such as oxidation and disulfide exchange (Manning et al. 1989). Lyophilized proteins have additional stability considerations (FDA 1993c). Protein degradations affect the pharmacology and availability of a drug, and FDA investigators are likely to review stability data from this perspective.

Stability data must support the potency and purity of a drug product during clinical trials; if trials last six months, for example, then the applicant must have six months of stability data. During process development, additional data will be generated identifying critical process parameters affecting stability for each stage; for many processes, these factors include temperature, pH, metal ions, proteases, air interfaces (bubbles and foam), and time.

Investigators will focus on finished product stability. For lyophilized products, critical process attributes include moisture residue, reconstitution properties, and assay. Reconstitution studies should include worst-case conditions; the product should be reconstituted at maximum and minimum concentrations, and moisture specifications should bracket the values that have yielded acceptable reconstitution results.

Identity

A product's label must accurately reflect the identity of the contents. In addition, commercial and clinical products require advisory statements. A product failing to display a required statement is misbranded, and product labeled inconsistently with its contents is adulterated and misbranded. The two categories of label errors are label mix-ups and product mix-ups. Product mix-ups are less frequent than label mix-ups; product mix-ups occur when a product is substituted during processing. Label mix-ups occur when the product is correct, but the label does not correspond with the product. The distinction depends on what the manufacturer was intending to produce.

Label Mix-Ups

The FDA remains intensely interested in packaging and labeling operations. It recently amended the GMP regulations (FR 1993; effective August 3, 1994) to tighten the controls on labeling operations. The new regulations allow manufacturers to use 100 percent electronic inspection in lieu of label reconciliation. In addition, the FDA narrowed the allowances for gang printing (labeling derived from a sheet of material on which more than one item of labeling is printed) (CFR 1994a). There is an additional requirement to identify filled, unlabeled containers ("bright stock").

Manufacturers of clinical supplies must abide by the GMP requirements. In response to a comment, the FDA stated in the *Federal Register:*

The revised labeling control provisions apply to the preparation of dosage forms that are under clinical investigation, whatever the size of the product lot. The small size of a product lot does not reduce the need for labeling controls. In fact, in some cases, the manufacture of many small lots may increase the opportunities for mix-ups and the need for label controls. Furthermore, in cases where firms label both investigational and noninvestigational products, the suggested exemption for investigational lots could create additional difficulties in labeling control for all products on the line (FR 1993).

Labeling refers to any written, printed, or graphic material on or accompanying the drug product, including shelf cartons, shipping cartons (if the drug name appears on them), and inserts.

During FDA inspections, investigators may examine the label storage area. Access to the area should be strictly limited. Labels should be stored neatly, separated from other materials. Investigators will look for outdated or discontinued labels in storage.

Many manufacturers enlist other firms to label their products. Product approval usually requires FDA inspection of these ancillary facilities. The applicant is ultimately responsible for GMP compliance at these facilities, and thorough auditing is prudent.

Product Mix-Ups

In some small biotechnology firms, investigators have found finished products stored loosely in refrigerators, freely mixed with unlabeled products, reagents, and laboratory samples. Such storage conditions welcome contamination from spills. Far more significantly, it is very easy to dispense the wrong material. Investigators may ask to see product storage before and after labeling operations.

Product identity is a principal concern for unique or customized products, for example, gene therapies, including engineered viruses, and somatic cell therapies. For some of these technologies, product mix-up can kill a patient. Accordingly, the FDA will continue to look closely at segregation, in-process identification, and final identification testing.

Each unit of a customized product should be positively identified (e.g., a written label on a culture vessel). Identification of the in-process material should be checked against the specified identification described in the batch record; these checks require documentation and verification.

Firms must show that final identification tests will distinguish among all of the products in the facility. Identification tests (e.g., modified Southern blots) must be robust and specific. Statistical arguments are persuasive, but a database of the products in the facility must show that each product is unique according to the battery of identification tests applied. These identification tests should be validated for ruggedness and selectivity. Investigators may audit identification tests performed for lot clearance to ensure that the data observed are clear and validated.

STRATEGIES FOR COMPLIANCE

The list of common problem areas in the first section of this chapter enumerated obvious deficiencies. In many cases, the citation reflected a complete absence of an attempt at compliance. Accordingly, the specific enumeration of expectations does not provide a comprehensive compliance strategy. The "current" Good Manufacturing Practice (cGMP) standard is based on what many would consider common sense; when the FDA raises issues with companies, there are generally few disagreements regarding the technical merits of the FDA's position. One can obviously ask, "Why does the FDA continually note relatively straightforward problems in an industry of undisputed scientific prowess?"

The answer to this question may rest with the entrepreneurial nature of the industry. Small start-ups may lack the institutional experience for completing major projects, and there may be some tendency to underestimate the magnitude of the responsibility. When budgets are tight and deadlines are short, senior managers may risk a minimal QC investment. Some firms view QC as an unavoidable chore rather than as an integral part of the process. Firms with excellent compliance histories invariably invest heavily in these six strategies for compliance: planning, review, auditing, management involvement, corporate culture, and thorough investigations of problems. These firms do not react to FDA expectations as much as they actively promote a continuing culture of quality.

Importance of Planning

As discussed, many problems at biotechnology firms are fundamental; many observations by investigators cite a complete lack of a quality system. Since most of the industry wants to be in compliance, it seems likely that some quality programs simply fall through the cracks. Neglect is understandable when one considers the labyrinth of scientific, financial, personal, medical, commercial, and regulatory considerations that a start-up company faces to bring a product to market. QA managers need to develop, execute, and review programs for validation, stability, training, equipment qualification, documentation, and analytical methods. No individual can organize such programs extemporaneously, particularly when the systems and their operational requirements are continually changing.

While there is no GMP requirement for a validation master plan, it is prudent to have one. A detailed and comprehensive validation master plan organizes the tasks involved in producing a biopharmaceutical. A master plan sets broad policies for every operational area. The master plan is a relatively new concept; while expectations are changing, some consultants have discussed their structure and contents (Hughes 1994).

From the FDA's perspective, the completed validation master plan can limit the time of the inspection by rapidly conveying management's systematic approach to compliance. If a firm lacks a validation master plan, an investigator must specifically check each critical quality system. The validation master plan can persuade an investigator that the firm has organized their internal compliance mechanisms. The plan can provide a checklist that an investigator can spot-check.

Importance of Review

Some firms review data only on an exceptional basis. Frequently, no one checks the original laboratory worksheets if the reported results comply with specifications. This approach does not meet the FDA's regulations or expectations. Areas with continuing review problems include environmental monitoring, stability, and component testing. In these areas, a small amount of exceptional data may get lost in the large volume of passing results. This tendency extends to voluminous documents such as validation reports, development reports, protocols, and lengthy batch records.

At one firm, the first lot made with a new manufacturing record contained 12 errors, including specifying equipment that the firm did not own and a calculation error in the percent yield calculations. When interviewed by the FDA, all 8 people who signed the master formula said that they looked at "everything" before signing off on the master formula.

In another case, a qualification run for a sterile process failed because the firm had set the specifications far too strictly. Explaining the failure to the FDA was extremely difficult and time-consuming. Ultimately, it was found that no one had actually reviewed the specifications in the protocol.

During batch record review, investigators are looking for unexplained deviations, significant product losses, calculation errors and other inconsistencies. Rare deviations are expected, but the FDA becomes concerned when an internal review does not detect common errors. Investigators sometimes find informal notebooks kept by engineers, analysts, and scientists; for clinical and commercial lots, these notes must comply with the GMPs, and the QC unit must review these notes (CFR 1994). If these records conflict with the official records, there may be serious regulatory concern.

A written procedure outlining how documents are to be reviewed might include responsibility, time frames, documentation and correction of problems, and, most important, the scope of the review. A validation master plan, if one is prepared, might address the review of validation data as they are compiled into a report.

Importance of Audits

People who work in the biopharmaceutical industry attend conferences, conduct training, and read articles (and book chapters) to deduce the FDA's conduct on inspections. In quality firms, FDA inspections cover only a small fraction of the material reviewed during internal audits. Internal auditors have a greater knowledge of a firm's problem areas than the FDA. In addition, internal auditors understand the science of a process better than the FDA. Finally, internal audits are ultimately constructive rather than adversarial.

Some audits are planned with the people who are being audited. Other audits are unannounced. A third variation of audits follows a specific complaint or request from an employee. These audit procedures should be formal. In an unfortunately large number of cases, firms have not pursued problems uncovered during audits until the FDA became involved. In one case, a QA vice president personally conducted an inspection at a vendor of a critical component. Despite documenting serious problems with the vendor's QC systems, this official nonetheless approved the vendor. Ultimately, the defective component led to stability failures and a recall.

In another firm, managers noted problems with an analyst's work quality, including submitting unsubstantiated results, during three distinct reviews, yet the managers allowed this employee to continue the objectionable behavior for several years. At this firm, each reviewer thought that another part of the organization had addressed the problem. To avoid situations of this nature, the most senior managers should be aware of the general nature of the audits, the findings, the corrective action plan, and the action taken. With this information, the senior managers can be assured that corrective action had been conducted according to the commitments. A senior manager who knows of and allows unacceptable conduct may be subject to criminal prosecution.

Senior managers should conduct some audits. When senior managers personally conduct internal audits, they are essentially auditing both production and the previous auditors. These audits prepare all employees for an FDA inspection. Employees also learn that senior managers have an active interest in their performance. Further, audits help employees to become confident in their understanding of procedures; employees who can intelligently explain their jobs raise the confidence of investigators. Investigators prefer to speak directly with the people who are responsible for a particular operation. When management (or regulatory affairs) shields an employee from the FDA, it raises suspicions about a manager's confidence in that employee's qualifications.

In general, the FDA will not request the results of internal audits because of the chilling effect such an external review may have on the audits. However, if there is evidence of a problem at a vendor, contractor, or supplier, the FDA may request the findings from those audits. Contract labs and manufacturers are as diverse as their customer firms. Many are excellent, but some are struggling with compliance. Nearly all applicants audit contract manufacturers, including manufacturers of critical components. For critical quality tests, contract labs should be audited as well. Some applicants elect to send spiked samples to contract labs to evaluate their test findings. Unfortunately, a few applicants accept data from the contract lab at face value; however, there should be documented review of the lab's procedures and a report. The applicant is responsible for all testing, whether performed internally or on contract. The FDA's procedures for product approval require a recent favorable inspection at every facility named in an application. Often, problems at an ancillary facility delay approval of an entire application.

Management Involvement

Recently, the FDA has enjoined major pharmaceutical companies for deficiencies in the way in which management has responded to continuing problems. FDA investigators continue to look closely at managerial oversight.

The FDA is concerned not only with the content of reports but also with the routing and distribution of these reports. For example, development reports (if they are prepared) should be routed through senior managers. In some firms, weekly data summaries of routine tests, such as environmental monitoring, circulate among the senior QC management. There is a GMP requirement for responsible officials to review complaints and other investigations; generally, the FDA wants to be sure that operations are monitored by the senior managers on whom legal responsibility rests (*U.S. v. Park* 1975).

Even at early stages of process development, senior managers should be aware of obvious quality milestones. Formal mechanisms for informing senior managers might include electronic routing lists and formalized executive summaries. Investigators become concerned when large quantities of significant data go to file without meaningful review.

Investigators may review executive summaries to determine whether senior managers are being kept abreast of significant manufacturing trends and problems. The FDA may also request these records to evaluate management's follow-up of failures. The FDA's inspectional authority extends to these records when they contribute to the written follow-up to investigations conducted in accordance with 21 CFR §211.192.

Corporate Culture

Quality is designed throughout the organization. The most important determinant of quality is the involvement of trained employees.

Recently, at a prescription drug manufacturer, production employees were found not recording in-process data. If a failing result was obtained on a particular computerized test apparatus, the analyst signaled the process operator to subtly adjust the machine and take a second sample. They printed only the passing result from the computer. During interviews with employees, the FDA found that production managers were telling employees that their jobs depended on increased production. In addition, there were cash incentives for the shift with the highest yield. Whether or not upper management was aware of the problem, they had created an environment in which the number of units produced became more important than a quality product.

In another firm, employees wrote tare weights on scrap sheets of paper, which the lead operator described as an "extra quality check" so the operators knew that they had weighed out the right amount "before QA got a look at it." In another case, the master formula specified a component in kilograms, but it was dispensed to the line in gallons. Employees calculated the conversion for gallons to liters to kilograms on the cardboard box that contained the bottles.

In another firm, a laboratory analyst found extreme inconsistencies in a coworker's work. When the analysis raised the concern to management, they ordered her to investigate and confront her coworker. Furthermore, the analyst had to defend the inadmissibility of the data to the most senior management in the firm. Under such circumstances, many employees feel pressured to ignore the problems rather than bring them to management's attention.

These are examples of a poor corporate culture. In the examples cited, employees were unaware of GMP concepts or afraid of QA. Employees need to feel free to suggest improvements to a procedure, method, or formula. There should be regular communication with the people who are actually doing the work so that managers recognize any need for change.

Importance of Investigations

In the much-cited injunction *U.S. v. Barr Laboratories*, the District Court for Newark, New Jersey, established guidelines for conducting investigations of out-of-specification results (Paulson 1994). Although the decisions of one district court do not bind courts in other districts, courts typically draw on particularly well-reasoned cases in other districts. The following discussion of investigations draws heavily on the *Barr* decision; thus, it is based on policy, not regulation or statute. Essentially, the FDA expects a firm to treat all out-of-specification results as valid unless there is clear, affirmative evidence to discount the noncompliant result. Sometimes, out-of-specification results cause a firm to reject a batch or a stability study.

Many firms have responded to the *Barr* decision by refining their procedures for conducting investigations of out-of-specification results. There have still been notable problems in some of these refined investigations, however.

- *No investigation of a failure to meet control limits.* Manufacturers set control limits to comply with the GMP requirements for scientifically and statistically based specifications. A control limit failure need not necessarily result in rejection of a batch, but it should trigger an investigation. This is not a burdensome expectation. If alert limits are set at 2 standard deviations, for example, then any excursions should occur by chance no more than 5 percent of the time; control limits based on 3 standard deviations should fail by chance significantly less than 1 percent of the time. Interpretation of these control limits should be based on standard statistical practices (Grant and Leavenworth 1980; Bolton 1990).

- *Acceptance of the first plausible explanation.* In a large number of investigations, the quality group grasps the first plausible explanation for the deviant observation. This situation would be justified if the purpose of the investigation were to explain the data away; however, the goal is to discover the reason behind the observation. One traditional pharmaceutical firm concluded that metal fragments in tablets had come from the brass feed frame. FDA analysis found that the metal fragments were not made of brass. In another case, a firm found that certain stability samples of a gel had low ethanol values. The firm blamed the failures on evaporation during delays in manufacturing. FDA review found that the container closure system had microscopic flaws that the firm's package integrity test did not detect. It is insufficient to find an initial single plausible reason for a failure. Alternatives must be considered and ruled out.

- *No written procedure for conducting investigations.* To ensure that investigations are timely, scientifically valid, and documented, a written procedure should contain the following, at a minimum.

    - **Responsibility:** There are inherent conflicts in assigning coworkers to inspect each other. A written procedure should clearly outline the responsibility for conducting investigations. In the best case, the investigator is impartial.

    - **Methodology/documentation:** If the written procedure for conducting an investigation includes a checklist, the procedure can be self-documenting. Even if the investigation cannot reach a conclusion, it is essential to document what alternatives the investigating team has considered. Documentation also helps managers track the status of major investigations.

    - **Time frames:** The FDA's *Guide to Inspections of Pharmaceutical Laboratories* (1993d) says that investigations should be concluded within 20 days. However, investigations should commence immediately. If the laboratory waits for 19 days to begin an investigation, the chance that the error will be identified decreases dramatically. Investigations should include reviewing the equipment settings, sample identification, and interviewing the operator or analyst. If a firm cannot clearly point to a laboratory error, they will need to greatly expand the scope of the investigation.

- **Management review/approval:** Managers should ensure that investigational conclusions are scientifically valid. Managers should also make sure that any necessary follow-up is conducted according to an established schedule. Managers should closely review inconclusive investigations to be sure that all avenues were explored. There is a possibility that the investigation may be conducted in a cursory manner when the people conducting the investigation have direct responsibility or authority for its outcome.

- **Trend results:** Finally, unusual findings should be trended concisely and reviewed. In a recent inspection, an investigator was looking at personnel monitoring data for the operators performing the final filling of the drug. There were periodic excursions, but no real pattern developed until the investigator sorted the data by operator name. It turned out that nearly all of the excursions were from a single operator. Over half the time that person sat at the hood, the results were "too numerous to count." Senior managers should trend problems to determine the need for systematic corrections, such as retraining or modifications in a procedure.

CONCLUSION

In addition to GMP inspections, the FDA has many other inspectional roles: for example, affirming the authenticity of submissions, investigating the promotion of prescription medications, and auditing a firm's response to complaints and injuries. This chapter provided a flavor of an FDA inspection by emphasizing the specific risks FDA is trying to minimize. Investigators should not be auditing firms for rote adherence to policies and regulations. Investigators should concentrate on significant, product-related manufacturing controls.

The problem areas in biopharmaceutical plants are generally fundamental. Most biopharmaceutical companies can draw on highly trained scientists to resolve the more complex issues. The basic nature of problems may be the result of inadequate managerial controls or a misunderstanding of FDA expectations.

The FDA does not provide a cookbook for compliance. Neither does it conduct all-inclusive inspections. Its role is to audit a company's control procedures. Accordingly, this chapter provided a general idea of why the FDA expects manufacturers to enact particular controls. When a company's quality commitment exceeds FDA's expectations, approvals are quick and regulatory headaches small.

ADDENDUM

Since this chapter was originally written for publication, the FDA has dramatically changed the administration of the biotechnology inspection program. CBER has harmonized its application, now called a Biologics License Application, with CDER's New Drug Application; this change has increased the importance and depth of inspections because less information is submitted in the application for review. After piloting highly focused inspection teams in two biotechnology-intensive regions, the FDA has expanded the concept to the "Team Biologics" initiative. This new program promises to bring highly trained investigators to all inspections of biologics and also greatly increases enforcement of GMPs.

The FDA continues to release their own guidance as well as participate in the ICH. Significant new regulations for electronic signatures and finished drug cGMPs (currently under review) codify many existing implied requirements as well as add additional responsibilities.

Despite these and other significant changes in the logistics of the inspection program, the inspection coverage and philosophy remain similar to the description in this chapter. Some of the industry's common and basic deficiencies have improved with time, but the general principle remains this: Companies continue to have difficulty with basic GMP compliance. For this reason, the strategies for compliance discussed in this chapter—planning, reviewing, auditing, involving management, and conducting thorough investigations of problems—remain timely.

ACKNOWLEDGMENTS

I am grateful to Dr. Ann Blake, Ruth H. Johnson, and Ilene D. Leahy-Wells for their expert technical and editorial advice. I am also indebted to the Division of Field Investigations for shepherding the manuscript through the review process.

REFERENCES

Adner, N., and G. Sofer. 1994. Biotechnology product validation, part 3: Chromatography cleaning validation. *BioPharm* 7 (3):44–48.

Avis, K. E., and M. J. Akers. 1986. Sterilization. In The theory and practice of industrial pharmacy, 3rd ed., edited by L. Lachman, J. L. Kanig, and H. A. Lieberman. Malvern, Penn., USA: Lea and Febiger.

Barry, A. R., and R. Chojnacki. 1994. Biotechnology product validation, part 8, Chromatography media and column qualification. *BioPharm* 7 (8):43–47.

Bolton, S. 1990. *Pharmaceutical statistics: Practical and clinical applications,* 2nd ed. New York: Marcel Dekker, Inc.

CFR. 1994a. Code of Federal Regulations, Title 21, Part 210, Good manufacturing practices in manufacturing, processing, packing, or holding of drugs; general. Washington, D.C.: U.S. Government Printing Office.

CFR. 1994b. Code of Federal Regulations, Title 21, Part 211, Good manufacturing practices for finished pharmaceuticals. Washington, D.C.: U.S. Government Printing Office.

Diers, I. V., E. Rasmussen, P. H. Larsen, and I-L. Kjaersig. 1991. Yeast fermentation processes for insulin production. In *Drug biotechnology regulation,* edited by Y. H. Chiu and J. L. Gueriguian. New York: Marcel Dekker.

FDA. 1983. *Guide to inspections of computerized systems in drug processing.* Rockville, Md., USA: Food and Drug Administration, Center for Drugs and Biologics.

FDA. 1987a. *Guideline on sterile products produced by aseptic processing.* Rockville, Md., USA: Food and Drug Administration, Center for Drug Evaluation and Research.

FDA. 1987b. *Guideline for submitting documentation for the stability of human drugs and biologics.* Rockville, Md., USA: Food and Drug Administration, Center for Drugs and Biologics.

FDA. 1987c. *Guideline on validation of the limulus amebocyte lysate test as an end-product endotoxin test for human and animal parenteral drugs, biological products, and medical devices.* Rockville, Md., USA: Food and Drug Administration, Center for Drug Evaluation and Research.

FDA. 1991a. *Guideline on the preparation of investigational new drug products (human and animal).* Rockville, Md., USA: Food and Drug Administration, Center for Drug Evaluation and Research.

FDA. 1991b. *Biotechnology inspection guide.* Rockville, Md., USA: Food and Drug Administration, Office of Regulatory Affairs.

FDA. 1993a. *Guide to inspection of high purity water systems.* Rockville, Md., USA: Food and Drug Administration, Office of Regulatory Affairs.

FDA. 1993b. *Guide to inspections of dosage form drug manufacturers—CGMPRs.* Rockville, Md., USA: Food and Drug Administration, Office of Regulatory Affairs.

FDA 1993c. *Guide to inspections of lyophilization of parenterals.* Rockville, Md., USA: Food and Drug Administration, Office of Regulatory Affairs.

FDA. 1993d. *Guide to inspections of pharmaceutical quality control laboratories.* Rockville, Md., USA: Food and Drug Administration, Office of Regulatory Affairs.

FDA. 1993e. *Guide to inspections of validation of cleaning processes.* Rockville, Md., USA: Food and Drug Administration, Office of Regulatory Affairs.

FDA. 1993f. *Points to consider in the characterization of cell lines used to produce biologicals.* Rockville, Md., USA: Food and Drug Administration, Center for Biologics Evaluation and Research.

FDA. 1994. *Investigations operations manual.* Rockville, Md., USA: Food and Drug Administration, Office of Regulatory Affairs.

FR. 1978. *Federal Register* 43 (190):45209. See comment 49.

FR. 1993. *Federal Register* 58 (147):41348–41354; effective date amended in *Federal Register* 59 (147):39255–39256.

Garnick, R. L., M. J. Ross, and R. A. Baffi. 1991. Characterization of proteins from recombinant DNA manufacture. In *Drug biotechnology regulation,* edited by Y. H. Chiu and J. L. Gueriguian. New York: Marcel Dekker.

Grant, E. L., and R.S. Leavenworth. 1980. *Statistical quality control,* 5th ed. New York: McGraw-Hill, pp. 280–281.

Grun, J. B., E. M. White, and A. F. Sito. 1992. Viral removal/inactivation by purification of biopharmaceuticals. *BioPharm* 3 (10): 22–30.

Hughes, T. 1994. Integrated master planning for validation. In *Proceedings of the BioPharm Conference.* Eugene, Ore., USA: Advanstar Communications.

Jungbauer, A., H. P. Lettner, L. Guerrier, and E. Boschetti. 1994. Chemical sanitization in process chromatography, part 2: In situ treatment of packed columns and long-term stability of resins. *BioPharm* pp. 37–42.

Klegerman, R. E., and M. J. Groves. 1992. *Pharmaceutical biotechnology: Fundamentals and essentials.* Buffalo Grove, Ill., USA: Interpharm Press, Inc.

Manning, M. C., K. Patel, and R. T. Borchardt. 1989. Stability of protein pharmaceuticals. *Pharm. Res.* 6 (11):903–918.

Muth, W. L. 1991. Fermenters for prokaryotic cells. In *Drug biotechnology regulation,* edited by Y. H. Chiu and J. L. Gueriguian. New York: Marcel Dekker.

Nelson, K. L., and S. Geyer. 1991. Bioreactor and process design for large-scale mammalian cell culture manufacturing. In *Drug biotechnology regulation,* edited by Y. H. Chiu and J. L. Gueriguian. New York: Marcel Dekker.

Paulson, W., ed. 1994. Quality control Reports. *The Gold Sheet* 28 (8):1–12. Chevy Chase, Md., USA: F-D-C Reports, Inc.

Phillips, G. B., and M. O'Neill. 1990. Sterilization. In *Remington's Pharmaceutical Sciences,* 18th ed., edited by A. R. Gennaro. Easton, Penn., USA: Mack Publishing.

Roche, K. L., and R. V. Levy. 1992. Methods used to validate microporous membranes for the removal of mycoplasma. *BioPharm* 5 (3):22–33.

Sadana, A. 1989. Protein inactivation during downstream separation, part II: The parameters. *BioPharm* 2 (3):20–23.

Scheer, L. A., W. C. Steere, and C. M. Geisz. 1993. Temperature and volume effects on filter integrity tests. *Pharm. Tech.* 17 (2):22–32.

Schier, J. 1995. Current good manufacturing practice regulations for pharmaceutical laboratories. *LC-GC* 13 (6):474–479.

Seely, R. J., H. D. Wight, H. H. Fry, S. R. Rudge, and G. F. Slaff. 1994. Biotechnology product validation, part 7: Validation of chromatography resin useful life. *BioPharm* 7 (7):41–48.

Stadtman, E. R. 1990. Metal ion-catalyzed oxidation of proteins: Bio-chemical mechanism and biological consequences. *Free Rad. Biol. & Med.* 9:315–325.

USP. 1995. The United States Pharmacopeia, 23rd ed. Taunton, Mass., USA: Rand McNally.

U.S. v. Park. 1975. *U.S. v. Park,* 421 U.S. 658 (6/9/75), reversing 499 F. 2d 839 (C.A.4, 1974) KKW 69-74 at 313.

Wang, Y. J., and Y. W. Chien. 1984. *Sterile pharmaceutical packaging: Compatibility and stability.* Philadelphia: Parenteral Drug Association.

Wechsler, J. 1994. Approving biologics or "this won't hurt a bit." *Appl. Clin. Trials* 3 (10):26–34.

White, E. M., J. B. Grun, C-S. Sun, and A. F. Sito. 1991. Process validation for virus removal and inactivation. *BioPharm* 4 (5):34–39.

5

VALIDATION OF BIOPHARMACEUTICAL PROCESSES

Howard L. Levine

BioProcess Technology Consultants

Francisco J. Castillo

Berlex Biosciences

Biotechnology in the pharmaceutical industry combines classical biochemistry and microbiology, traditional pharmaceutical technology, and biochemical engineering with advances in genetic engineering and cell fusion technology to prepare proteins and other biologic products for testing as potential therapeutic or diagnostic agents. These biologics are produced in a suitable host cell by fermentation or cell culture processes. Cells from a qualified cell bank are expanded in culture until sufficient numbers are obtained for the desired production scale. During culture, the product of interest may be either retained intracellularly or secreted into the culture medium. If the product remains inside the cell, then the cells must be harvested and disrupted, and the debris that is created removed to yield a particulate-free extract for further purification. If the product is secreted, the cells must be separated from the conditioned culture medium prior to purification. These steps are shown schematically in Figure 5.1. Following this initial product isolation,

Figure 5.1. Generic scheme for biopharmaceutical manufacturing.

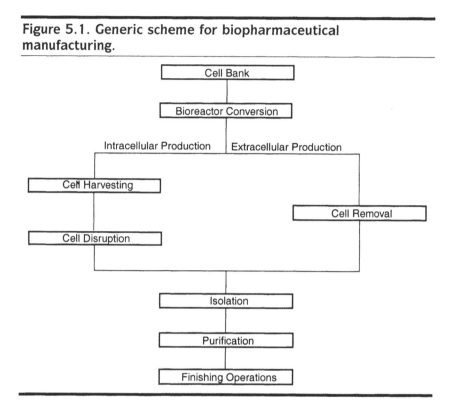

purification is accomplished through a combination of chromatography, filtration, and ultrafiltration.

The goal of validation is to demonstrate that a process, when operated within established limits, produces a product of appropriate and consistent quality. While successful validation may involve an inordinate amount of work over a long period of time and consume scarce resources, validation of critical process parameters is vital from both a quality and business perspective. During validation, the critical process parameters should be identified; based on sound scientific principles, appropriate studies should be performed to demonstrate that the parameters can be met on a consistent basis.

Validation is a scientifically rigorous and well-documented study that demonstrates that a process or piece of equipment consistently does what it is intended to do. Due to the complex nature of proteins and the relatively short histories of some cell lines in pharmaceutical manufacturing, it is difficult to characterize a biologic product fully. Thus, final product testing alone is insufficient to ensure the consistent manufacture of these products. Therefore, the processes used for the purification of proteins must be designed and validated to remove potential contaminants. The contaminants, which may arise from source material, equipment, or purification reagents, can include endotoxins, viruses, nucleic acids, and proteins, as well as media constituents, process chemicals, ligands leached from chromatography media, and modifications or inactive forms of the protein itself.

Validation should be considered as early in the development of a process as is practical. In this way, data required for validation can be collected during development studies and the production of batches for clinical studies. Valuable validation data can be collected during the production of products for Phase III clinical trials. In addition to in-process testing, the evaluation of products in humans under carefully monitored clinical trials provides the ultimate test of the safety and efficacy of the product.

Before validation is begun, the biologic product should be defined in terms of its physical and biological characteristics (Jones and O'Connor 1985). Recent guidelines from the U.S. Food and Drug Administration (FDA) have described the types of information that are considered valuable in assessing the structure of recombinant deoxyribonucleic acid (r-DNA)–derived proteins (FDA 1985, 1992, 1993, 1996b and c; FR 1996). Once the product has been fully characterized, specifications should be established to ensure uniformity, and the required level of purity should be established based on the indicated use of the product. Assays used to determine product purity should be validated to ensure that the sensitivity of the analytical test methods permit accurate detection and quantitation of the product as well as impurities. Once these criteria are met, validation of the manufacturing process can begin.

The process validation of biopharmaceutical processes usually entails four major areas:

1. Cell bank testing and qualification.

2. Process chemicals and raw materials, including media components for fermentation or cell culture, column packing materials used in chromatography, and membranes used for tangential flow filtration.

3. Equipment qualifications.

4. Performance qualification of the process itself.

Equipment qualifications are normally broken down into installation qualification (IQ), and operational qualification (OQ), and performance qualification (PQ). IQ and OQ ensure that the equipment is properly installed, calibrated, and functioning according to specifications, while PQ establishes that the equipment can reliably perform the function for which it is to be used.

The PQ of the manufacturing process establishes that the process is effective and reproducible and that the final product meets all established release specifications. In process validation, it is important that protocols clearly specify the procedures and tests to be conducted, the data to be collected, and the acceptance criteria. The purpose for which data are collected must be clear, the data must reflect facts, and the data must be collected carefully and accurately. The protocol should also specify a sufficient number of replicate process runs to demonstrate reproducibility and to provide an accurate measure of variability among successive runs. All important process variables should be identified, monitored, and documented. Analysis of the data collected establishes the variability of process parameters for individual runs and whether or not the equipment and process controls are adequate to provide assurance that product specifications will be met.

CELL BANK TESTING AND QUALIFICATION

Cell Banks

Biopharmaceutical products, including hormones, growth factors, cytokines, recombinant vaccines, monoclonal antibodies, and others in development, are made by expression in cells, either microbial (*Escherichia coli,* yeasts), insect, or mammalian (Chinese hamster ovary [CHO] cells, murine myelomas, hybridomas), which are grown utilizing a variety of validated techniques (Holmer 1997; Mossinghoff 1996).

The earliest validation in any biopharmaceutical process addresses the qualification of the expression system. The choice of an expression system for the production of heterologous proteins intended as pharmaceuticals is a very important step (Castillo 1995). Following gene transfer or cell fusion, considerable effort goes into selecting the producing cell. This includes corroboration of acceptable production yields, monoclonality, correct product characteristics, stability of productivity, and freedom from contaminating agents, such as microorganisms and viruses. This work is done to guarantee the availability and quality of the cell expression system and its capability for generating the desired product in a consistent, cost-effective manner.

To protect the results of this costly effort, it is important to maintain and preserve the cells to ensure permanent access to viable, homogeneous, and axenic stock cultures. Preservation also minimizes the potential for genetic changes, which may lead to reduction or loss of desired phenotypic properties, including decrease in specific productivity. Techniques for the maintenance of microorganisms are varied and include storage under paraffin, desiccation, lyophilization, and cryopreservation (Chang and Elander 1986; Kirsop 1987, 1991; Gherna 1981; Onions 1983; Calcott 1996; Smith 1988; Glazer and Nikaido 1995). Animal cell stocks are always maintained frozen.

For manufacturing purposes, the FDA recommends the generation of a cell banking system to ensure an adequate supply of equivalent cells for use over the entire life span of the product. The cell bank system should consist of two tiers: a master cell bank (MCB) and a manufacturer's working cell bank (MWCB).

Master Cell Banks

The FDA defines an MCB as "a collection of cells of uniform composition derived from a single tissue or cell, and stored in aliquots under defined conditions (FDA 1985, 1992, 1993). The MCB is intended for the preparation of MWCBs, but often it is also used, during the initial phases of product development, for the preparation of material needed for purification and formulation development work, preclinical and toxicology studies, and initial medical trials.

Manufacturer's Working Cell Bank

MWCBs are derived from one or more vials of the MCB. Cells from the MCB are expanded by serial subculture up to a selected passage number, at which point the cells are combined into one pool, concentrated, and aliquots dispensed into individual ampoules or vials and stored under defined conditions (FDA 1985, 1992, 1993).

One or more vials from the MWCB are used to produce one lot of a biological product.

End-of-Production Cell Banks

When utilizing fermentation or cell culture for production, end-of-production cells (EPCs) must be tested at least once to evaluate whether new contaminants are introduced or induced by the growth conditions. Changes in the growth medium or the scale of operation also require testing of EPCs to determine their impact on product yields and consistency. To ensure their stability, banks of EPCs should be prepared and stored under the same conditions as the MCB and MWCB.

Preparation of Cell Banks

The successful preparation and maintenance of cell banks are key activities in establishing a process to produce a biologic. Even with experienced operators, these activities require important preliminary work to determine how the cells respond to freezing, how to select the best conditions for freezing, and how to execute freezing. Several principles apply in general for the preparation of cell banks to ensure consistently good recoverability when the cells are thawed. A very important one is the use of healthy, exponentially growing cultures. Detailed protocols are described in the literature (Castillo 1995; Coriell 1979; Daggett and Simione 1987; Glazer and Nikaido 1995; Hay 1989; Wiebe and May 1990). For most mammalian cells, the following practices and precautions have been found to work well:

- Centrifuge the cells and resuspend at high concentration (i.e. $\geq 10^7$ viable cells/mL) using fresh, cold (2–8°C) medium to slow cell metabolism and minimize substrate utilization and metabolite accumulation.

- Work fast, as a concentrated cell suspension will quickly deplete nutrients, accumulate waste, and acidify the medium, all of which negatively impact the cells and the outcome of the freeze.

- Divide the cell suspension in aliquots and thoroughly mix one at a time with fresh medium containing sufficient dimethylsulfoxide (DMSO) to provide adequate concentration for cryopreservation (usually between 5 and 10 percent DMSO). Quickly dispense into vials and transfer to a programmable freezer or corresponding places in the vapor phase of a liquid nitrogen freezer. Avoid delays and proceed to the next aliquots until all are used.

- Store the vials of all cell banks in more than one location to prevent complete losses due to freezer failure.

Testing of Cell Banks

Irrespective of the expression system, testing of the cell banks is required to ensure identity, consistent performance, and freedom from contamination. The types and number of tests will depend on the species and origin of the cell bank, the manufacturing process, and the intended application. As a rule, the most extensive testing is done on the MCB. Most of the tests can be contracted to one or more of the service companies listed in Table 5.1. Tables 5.2 to 5.4 summarize the tests recommended by the FDA for different banks.

Table 5.1. Companies That Provide Contract Testing Services for Cell and Virus Banks

Company	Address
Genzyme Transgenics (GTC Washington Laboratories)	Two Taft Court Rockville, MD 20850 Tel: 888-448-2522 Fax: 301-738-1061
Inveresk Research International Ltd.	Musselburgh EH21 7UB, Scotland, UK Tel: 031-665-6881 Fax: 031-665-9976
MA Bioservices, Inc. (Formerly Microbiological Associates)	9900 Blackwell Road Rockville, MD 20850 Tel: 800-553-5372 Fax: 301-738-1036
Panlabs Biosafety	11804 North Creek Parkway South Bothell, WA 98011 Tel: 800-487-9161 Fax: 206-487-3787
Quality Biotech	1667 Davis Street Camden, NJ 08104 Tel: 800-622-8820 Fax: 609-342-8078
Tektagen	358 Technology Drive Malvern, PA 19355 Tel: 800-648-6682 Fax: 215-889-9028
TSI Corporation	5516 Nicholson Lane Kensington, MD 20895 Tel: 703-642-5671 Fax: 703-642-5691

Virus Banks for Gene Therapy

Gene therapy represents a collection of strategies for the treatment of human disease based on the transfer of genetic material (DNA) into individuals. The goal is to modify the phenotype of the targeted cell through alterations of its genotype. Genes can be transferred in

Table 5.2. Testing of Microbial Banks

Test	MCB	MWCB	EPC[1]
Identity tests			
Restriction maps	+	+	+
Vector sequence			
Phenotypic markers			
Auxotrophy, antibiotic resistance			
Stability			
Plasmid retention	+	+	+
Sterility			
Free of bacteria and fungi	+	+	+
Free of bacteriophages	+	+	+
Growth kinetics	+	+	+
Production kinetics	+	+	+

[1] Tests to be done at least once on EPC.

Source: FDA (1985, 1992)

two ways: (1) treating cells ex vivo for reintroduction into the patient or (2) treating the patient by direct, in vivo administration.

For the delivery of genes, viral and nonviral systems are used. To date, the majority of clinical protocols have used viral vectors (Blau and Khavari 1997; Crystal 1995; Verma and Somia 1997), among which retroviruses and adenoviruses are the most widely used (Orkin and Motulsky 1995; Ross et al. 1996). Retroviruses were the first vehicles developed for gene therapy and have the advantage of being stably integrated into the genome of dividing cells and, consequently, permit long-term gene expression.

Adenoviruses do not integrate into the genome and are commonly used when a high level of transient expression of the transgene is required. Adeno-associated virus (AAV) vectors are increasingly considered for treatments where long-term expression

Table 5.3. Tests for Hybridoma Banks

Test	MCB	MWCB	EPC
Identity tests			
Isoenzymes	+	+	+
DNA fingerprinting			
Growth kinetics and cell yields	+	+	+
Production kinetics and yields	+	+	+
Product quality	+	+	+
QC test battery			
Sterility	+	+	+
Free of bacteria and fungi			
Free of mycoplasma	+	+	+
Species-specific viruses	+	–	–
(i.e., mouse antibody production [MAP] assay for murine hybridomas)[1]			
Retroviruses	+[2]	–	+[2]
Infectivity			
Electron microscopy			
Reverse transcriptase			
Adventitious viruses	+	–	+
In vitro assay			
In vivo assay			

[1]Tests to be done at least once on EPC.

[1]Human cell banks are routinely tested for the following viruses: human immunodeficiency virus (HIV) types 1 and 2, cytomegalovirus (CMV), hepatitis B (HBV), Epstein-Barr virus (EBV), parvovirus B-19, herpes virus 6 (HHV-6), and human T-cell lymphotropic virus (HTLV) types I and II.

[2]Retrovirus testing not required for murine hybridomas.

Sources: FDA (1993, 1996b, 1997a)

Table 5.4. Tests for Mammalian Cells Banks

Test	MCB	MWCB	EPC
Identity tests			
Isoenzymes	+	+	+
DNA fingerprinting			
Growth on soft agarose	+	−	−
Growth kinetics and cell yields	+	+	+
Production kinetics and yields	+	+	+
Sterility	+	+	+
Free of bacteria and fungi			
Free of mycoplasma	+	+	+
Species-specific viruses	+	+	+
(i.e., hamster antibody production [HAP] for CHO cells)[1]			
Retroviruses	+	−	+
Infectivity			
Electron microscopy			
Reverse transcriptase			
Virus contaminants	+	−	+
In vitro assay			
In vivo assay			

[1] Human cell banks are routinely tested for the following viruses: HIV types 1 and 2, CMV, HBV, EBV, parvovirus B-19, HHV-6, and HTLV types I and II.

Sources: FDA (1966, 1993, 1997a)

of genes is required because this virus integrates into the q arm of chromosome 19. This is an attractive property because, compared to retroviruses, the risk of insertional mutagenesis is minimized (Muzyczka 1992). Alternative vectors in development include herpes, *Vaccinia,* sindbis, rabies, influenza, and HIV (Jolly 1994).

While microbial and mammalian cells have been used safely for years in the production of biologics, the experience in gene therapy is limited, and the use of viral vectors raises safety concerns. To address these concerns the FDA and National Institutes of Health (NIH) have jointly sponsored the annual International Gene Therapy Conference since 1995. For additional information regarding these conferences, the reader may refer to the FDA website on the internet (GTINFO@A1.cber.fda.gov), or may call the FDA Division of Cellular and Gene Therapies at 301-827-0681. The FDA has also issued a "Points to Consider" document for human somatic cell and gene therapy (FDA 1996a).

The production of retrovirus vectors requires the establishment of a producing cell line that expresses the vector. This is done by transfecting cells with a plasmid containing the recombinant vector genome. The plasmid integrates randomly in the cell genome and directs production of the vector that is continuously released from the cells that remain viable. A stable cell clone is selected as the producing cell line, and cell banks prepared. Extensive testing of the cells and their supernatants includes, in addition to adventitious agents, the presence of replication-competent retroviruses (RCRs), which are tested for at the levels of MCB, MWCB, production supernatants, and postproduction cells (FDA 1996a).

Unlike retrovirus vectors, the preparation of adenovirus vectors requires two components for every lot—replication-incompetent virus seed and packaging cells—that provide the gene products required for vector multiplication upon infection. The most frequently used line is HEK 293, a human embryo kidney line transfected with the adenovirus 5 E1 gene (Graham et al. 1977; Aiello et al. 1979). The cells are grown, banked, and tested as described above.

Preparation and Testing of Adenovirus Stocks

The preparation of adenovirus vector banks is preceded by consecutive plaque purifications of the vector seeds to ensure monoclonality. Plaque purification is done using qualified packaging cells free of contaminants. Once the plaques have been purified, they are expanded to produce virus stocks that are then tested to verify the absence of bacteria, fungi, mycoplasma and replication-competent adenoviruses (RCAs).

Preparation and Testing of Adenovirus Banks

Once a qualified virus stock is selected, it is expanded as needed to generate material to prepare a master virus bank (MVB). Testing of the MVB is done at two levels: the crude harvest and the purified virus. The harvest is tested for the presence of bacteria, fungi, mycoplasma, inappropriate viruses (by in vivo and in vitro assays), and specific human viruses (see footnote in Table 5.4). The virus is purified from the harvest, filter sterilized, and dispensed into vials to create the MVB, which is then stored at $\leq -70°C$. Vials from the MVB are tested for the presence of RCA. The FDA recommends that adenovirus gene therapy products contain less than 1 percent RCA per dose (FDA 1996a).

Cellular and Tissue-Based Products

The FDA recently issued the first proposed approach to regulating human cellular and tissue-based products by defining criteria for product characterization (FDA 1997b). In this document, the FDA lists the recommended and required testing for these products. The most stringent apply to allogeneic viable tissues and include a quarantine until tests for specific communicable bacterial and viral diseases are completed.

EQUIPMENT QUALIFICATION

As part of the overall validation of biopharmaceutical processes, all equipment associated with the production of the biologic product should be qualified and validated. For fermentation or cell culture, process equipment includes the fermentor or bioreactor used to produce the crude harvest (Asenjo and Merchuk 1995), as well as incubators, seed fermentors, media preparation equipment, sterilization and delivery systems, biosafety hoods, and small instruments such as pH and conductivity meters, balances, and scales. For downstream processing, this equipment will include systems for chromatography, ultrafiltration, microfiltration, and other purification operations, and equipment used for buffer preparation, process monitoring, and product handling. To initiate equipment validation, all instruments must be properly calibrated to ensure their correct and accurate operation (Bremmer 1986), and biosafety hoods should be certified to ensure the integrity of the high efficiency particulate air (HEPA) filter and the proper circulation of air (Kruse et al. 1991). As with other equipment used in pharmaceutical processes, the validation of equipment used in biopharmaceutical processing involves IQ, OQ, and PQ (Naglak et al. 1994).

Installation Qualification

The IQ of process equipment for biopharmaceutical processing is documented verification that all aspects of the installation of the equipment adhere to the manufacturer's recommendations; appropriate federal, state, and local safety, fire, and plumbing codes; and approved company specifications and design intentions. IQ demonstrates that the user of the equipment has purchased and installed the right equipment for the specific task. This document demonstrates that the user has considered aspects of compatibility of the equipment with the process and that the user has Standard Operating Procedures (SOPs) for keeping the equipment calibrated and in good operating condition through a preventive maintenance program and spare parts inventory. This document also demonstrates that the user has analyzed the operation of the equipment and determined the level of operator training required by preparing written SOPs covering these activities. Process equipment IQs should contain the following information:

System Application

The "System Application" section should briefly describe what processes are to be performed and where the equipment is located. As an overview of the system, a schematic diagram of the system is essential to a complete understanding and description of the system. Additionally, the equipment summary may also include design criteria for the equipment.

Equipment Information Summary

A detailed description of the system, including an equipment summary (manufacturer, model number, serial number) and a description of the components, should be provided. Each component of the system should be listed and described separately, with sufficient information to define the system clearly. For example, in a chromatography system, the equipment summary might include feed tanks, tubing or piping, pumps, filters, pressure gauges, valves, detectors, and the column itself. For tangential flow systems, such as those used for cell harvesting or product concentration, the equipment summary should describe the pumps, piping, instrumentation and controllers, the holding vessel, and the membrane (type, manufacturer, etc.).

Utility Description

All of the utilities supporting the process equipment should be described and checked to ensure proper installation. For example, the electrical source (voltage, amperage, etc.) should be listed and checked against local codes and the electrical specifications of the system. If the system requires compressed gases or steam, these utilities should be validated, and their quality and source should be described and verified.

Standard Operating Procedures, Manual Listing, and Other Documentation

The title and location of all appropriate manuals should be listed, and a checklist should be prepared to ensure that all the manuals exist and have been referenced in the installation of a piece of equipment. All SOPs relating to the installation, operation, and maintenance of the equipment should be listed. These documents should also contain the piping and instrumentation drawings (P&ID) and schematics necessary for installing, maintaining, and repairing the system.

Spare Parts and Service Requirements

A detailed list of recommended spare parts and their location is usually included in the IQ. This spare parts list may either be a separate list or included in the manuals. The IQ should also list and review maintenance procedures to provide assurance that prescribed maintenance can be performed without negative impact on the process or product.

Operating Logs

A listing of the name and location of logbooks that document the use of process equipment is usually included in the IQ document.

Process Instrumentation

The type, manufacturer, range, use, and calibration schedule of all process instrumentation should be listed. This list should be divided into critical and noncritical instruments. A *critical instrument* is one whose failure would adversely affect the product's quality or safety. Depending on system design and complexity, not all instruments are critical instruments. For example, if an ultrafiltration system is equipped with a flowmeter but process performance is not a strong function of flow rate, then the flowmeter may be considered a convenience or *noncritical instrument* in this system. The distinction is important because critical instruments will be calibrated and maintained on a more rigorous schedule than noncritical instruments. Also, change orders for a critical instrument will undergo a more extensive examination, and failure of a critical instrument during the process will be reviewed more carefully than failure of a noncritical instrument. All instrumentation on the process system should be calibrated against standards traceable or comparable to the National Institute of Standards and Technology (NIST). The IQ should also list the SOPs that describe the calibration procedures for these instruments.

Materials of Construction

Those items that come or may come in contact with the product should be described and verified to be compatible with the product and/or process. All components of the system, including lubricants with the potential for contacting the product, filters, valves, tanks, and so on should be included. Equipment vendors can often provide appropriate compatibility data; however, the user may have to confirm such data with actual process fluids. If materials leach from the system into the product stream, then it should be demonstrated that subsequent process steps remove these materials.

Operational Qualification

The OQ is documented verification that a piece of equipment or process system, when assembled and used according to the SOPs, does, in fact, perform its intended function. As with the IQ, the OQ is concerned with the equipment, not with the product or process per se. The OQ demonstrates that the user has tested the equipment and has found it to be free from mechanical or design defects before use in the production process.

Before starting the OQ for any process equipment system, the IQ on that system should be completed. Any required calibration for the system should also be completed. Calibration may be part of the IQ or OQ, depending on the format chosen by the organization. The IQ and OQ for supporting utilities, such as water systems, lighting, heating/cooling, and electrical, should also be completed prior to starting the OQ for the equipment system. The OQ document should include the following information:

SOP Audits

It should be verified that operators have received the proper training and are able to operate the equipment as intended by following the appropriate operating SOPs. If the equipment is automated, the tests should verify that the equipment responds to the controller as designed.

Checking Manual Elements

The manual elements of the system, such as hand-operated valves and traps, should be checked physically and/or visually to ensure proper operation.

System Integrity

The equipment should be tested to establish that it is capable of operating without leaks. The simplest means of detecting leaks is by visual inspection of the fluid path. Leaks may also be detected in complex systems by demonstrating that the fluid output equals fluid input (fluid mass balance). Pressure hold tests on the components and piping can identify leaks before attempting operation. Membrane manufacturers may be consulted to obtain recommended test procedures and specifications for verifying the integrity of filters once installed in systems. If the system is designed to provide biological containment of recombinant organisms, tests should be designed to address this issue prior to introduction of viable organisms to the system.

Flows/Pressures

Pumps should be tested to show that they deliver the required flow under normal operating conditions. Tolerances may be established for variations in flow rates.

Gradient Formation

For chromatography systems operating in a nonisocratic mode, the ability of the system to deliver reproducible gradients of the desired shape and slope should be demonstrated. The effect of variations in the formulation of individual solutions in the final gradient should also be determined.

Fraction Collector

If a fraction collector is used during chromatography, its correct functioning should be established. The accuracy of the timer/ volume counter and correct positioning of the delivery arm should be demonstrated.

Detectors/Recorders

If the data generated by detectors are to be used in process control, then the acceptable operating range, the limits of linearity response, the reaction time, and the response of the detectors and recorders with operating flow rates should be established.

Filters

Filters and filter housings should be examined to verify that they are appropriate for use with the flow rates and pressures likely to be encountered in the system. They should be suitable for their intended purpose, whether that be sterilization or particle removal. If filters are used for sterilization, then they should be validated as such.

Computer Control

If computer control is to be used in the operation or cleaning of the equipment system, validation of the control software and hardware in the system must be addressed (PMA 1986). It should be shown that the software functions correctly and is protected from unauthorized alteration. The ability of the system hardware to perform its assigned task should also be shown. A schematic of the control logic, including "if-then loop paths," should be included.

Alarms

All alarms should be tested by simulating "alarm conditions," either by actually challenging the system or by electronic simulation. For example, a pressure alarm may be tested by increasing the pressure in the system using pumps and valves; alternatively, the high pressure may be simulated by sending the appropriate voltage to the alarm mechanism.

Other Features/Components

Finally, each system may have unique features or components not found in conventional systems used for other applications. Appropriate tests to demonstrate the correct functioning of these features or components should be included in the OQ.

Performance Qualification

The PQ of process equipment will establish the reliable and reproducible performance of the equipment. Most aspects of equipment PQ are incorporated into process PQ. However, two important aspects of equipment PQ that should be performed before process validation begins are equipment cleaning (Brunkow et al. 1996) and sterilization (Leahy 1986).

Some bioreactors employed in continuous perfusion cultures, such as hollow-fiber systems, are constructed of disposable elements that are presterilized and preassembled. These bioreactor systems are dedicated for single use and discarded after decontamination.

VALIDATION OF FERMENTATION/ CELL CULTURE PROCESSES

A fermentation or cell culture process may involve microbial or animal cells and their growth in batch, fed-batch, semicontinuous, or continuous mode (Pirt 1975). In all cases, the process is run aseptically in a defined, controlled fashion, with the objective of producing a consistent harvest from which product can be recovered and purified.

Raw Material Validation

Raw materials are routinely tested to confirm their identity and purity. For materials used in cultures, the bioburden and endotoxin levels are also determined. Additional tests for culture media components include growth promotion, lack of growth inhibition, and promotion of product expression. Animal-derived components, such as sera, enzymes, and other proteins, are also tested for species specific viruses. Viruses, such as bovine viral diarrhea (BVDV), are frequent contaminants of sera and are often undetected in some of the assays employed by suppliers. This virus has a wide host range and can contaminate cell cultures (Bolin 1994). To eliminate the risk of infectious virus contaminants, the use of gamma-irradiated sera is recommended.

Process Validation

The goal of fermentation/cell culture process validation is to demonstrate product consistency, which, in itself, is a function of variables that include cell line stability, consistency of raw materials, and consistency of the culture environment/process (Seamon 1992).

Examples of process validation of batch and continuous processes for the production of recombinant proteins and monoclonal antibodies have been published (Castillo et al. 1994; Lubinecki et al. 1992; Wiebe and Builder 1994).

While flexibility in the approaches to process validation reflect the differences in recombinant products, the validation exercise requires the establishment of appropriate, duly validated assays, utilizing the most effective technology (Bebbington and Lambert 1994).

VALIDATION OF DOWNSTREAM PROCESSING

The purification of biotechnology products usually includes several chromatography and tangential flow (or cross-flow) filtration steps. Typically, the chromatography steps provide product purification; the tangential flow filtration steps are used for removing or concentrating whole cells or insoluble lysate components, concentrating macromolecules, buffer exchange, and depyrogenation. For most recombinant proteins, several chromatography steps are necessary to achieve the level of purity required for a protein to be used as a therapeutic agent. These purification processes are based on different properties of the biomolecule, such as shape, size, net ionic charge, hydrophobicity, or specific affinity to another biomolecule. Consequently, there are four main types of column-based separations: gel filtration or size exclusion, ion exchange, reverse-phase or hydrophobic, and affinity. These operations are normally carried out in a packed column connected to a pump to move fluids through the column, with a detector to monitor the effluent stream of the column. These processes are generally batch processes, and the packed columns are intended for repeated use.

Many, if not most, production-scale column separations are adsorption/desorption operations. In adsorption or on-off chromatography, the process stream is fed onto a column until the desired amount of product or impurities is bound to the medium. The product or impurity moves rapidly through the column until an unoccupied binding site is found where it adsorbs and remains until the elution conditions are changed. Weakly bound species are displaced by more tightly binding ones. After loading, unadsorbed or weakly bound material is washed from the column, and the desired species eluted by changing the elution conditions either gradually in a gradient or in a single step. In gel filtration, or size exclusion chromatography, molecules are continuously separated by their differential migration through a packed bed.

While validation at full production scale is preferred, cost, practical issues, and safety often make it impractical to generate all the necessary data. For these reasons, laboratory studies utilizing scaled-down columns and "spiking" experiments can yield acceptable validation data. These studies must be designed to model the production process, and product yields and purity must be comparable to those obtained in typical full production runs. Studies characterizing the physicochemical and biochemical interactions in the purification process may be performed at smaller scale. Also, clearance studies performed by challenge or spiking studies with radiolabeled chemicals, toxic chemicals, or infectious biological agents should be done at small scale to maintain worker safety and to avoid contamination of production equipment. Each process and validation study should be examined on a case-by-case basis to assess the legitimacy of scaled-down experiments.

Studies involving performance characteristics that are intrinsically dependent on equipment design cannot be modeled on scaled-down equipment and should be performed using the actual production equipment. The effectiveness of washing and cleaning cycles in the control of endotoxin and bioburden levels are examples of such performance characteristics. The equipment features affecting these characteristics involve the nature of fluid flow and the presence of void spaces in pumps, valves, flow adapters, and the column bed itself and, therefore, cannot be modeled in scaled-down experiments.

As in all scientific testing, it is important that validation tests and challenges be repeated enough times to ensure reliable and meaningful results. Demonstration at production scale that the manufacturing process consistently removes known and potential contaminants may eliminate the need for testing every production batch for certain impurities. Revalidation may be required if a process step is removed, added, or modified; when raw materials are obtained from new sources; or when process equipment is changed or modified.

Process Chemicals and Raw Materials

Chemical reagents, such as buffer salts, used to prepare solutions for downstream processing should be handled and controlled in the same manner as other raw materials used in pharmaceutical production. For example, appropriate raw material sampling plans and specifications should be developed and approved by quality control (QC) personnel. Test procedures and stability data should be developed and validated. Procedures should be developed to ensure that only approved materials that have been adequately sampled and tested are released for production use. In addition to chemical identity and purity specifications, raw material specifications should also include limits for levels of bacterial endotoxins and bioburden.

The water used in downstream processing should have consistent quality appropriate for the process because the type of water may influence the quality of the final product. The bacterial endotoxin and bioburden levels of the water used in downstream processing are important because they affect the level of these contaminants in the final product. In addition, the metal ion content of process water may be critical in some purification operations.

Water for Injection (WFI) is generally used for buffer preparation in downstream processing, especially for column-based separations. WFI should always be used for the final steps of a purification process. In any case, the water used in downstream processing should meet predefined specifications, and the water producing system itself should be properly validated.

Column Packing Materials

Chromatography media for column-based separations are selected to accomplish well-defined functions in the purification process and may have a significant effect on the purity, uniformity, and other characteristics of the product. The packing material should be treated like any other process raw material (i.e., quarantined on receipt and released for use only after meeting specified criteria). If the media supplier adheres to the Parenteral Drug Association's (PDA) vendor certification plan (PDA 1989), then a certificate of analysis specifying the results of physical, chemical, and functional tests of the medium, along with appropriate QC testing by the user may be all that is required on a lot-to-lot basis. However, end users should assess the vendor's release specifications and ensure that they are relevant and appropriate for their particular use. User tests might include, but are not limited to, pyrogen or bacterial endotoxin testing, particle size and distribution measurements, determination of the swelling factor for dry packing material in buffer, titration curves for ion exchange media, the separation of a standard protein mixture for size exclusion media, and binding capacity for all adsorption media. Most chromatography media will initially require a protein capacity test, ligand identification test, and backbone identification as part of raw material testing (Del Tito et al. 1996). Capacity testing can often be eliminated once historical data are analyzed, and vendor qualification is completed. In addition, chromatography media require particle size verification. Particle size can be affected by shipping and storage conditions and can be verified by particle distribution analysis (Del Tito et al. 1996) or in a functional way by using pressure flow curves for packed columns (Seely et al. 1994).

However, these simple tests may not always be sufficient to qualify a lot of chromatography medium for use in the production process. Therefore, it is essential that the user carefully analyze the process to determine if extra testing is required. Additional tests would then be applied to each incoming lot of medium. The most important criterion is that the packing material gives a specified purity and yield of the desired product. Other factors, such as selectivity or available capacity for the substance of interest should be defined within set limits. Specific details for testing various chromatography media may be obtained from the media manufacturer or the literature and will vary with the nature of the media (Dunn et al. 1991; Tsai et al. 1990).

In systems where the process batch size is small, or in highly automated and recycled systems, prepacked chromatography columns may be obtained from a vendor in a ready-to-use format. Prepacked columns present a special case since the contents and the product contact surfaces cannot be inspected by the user. In these cases, the user must have assurance in the form of a certificate of analysis from the manufacturer that the column meets previously agreed-on specifications with respect to materials of construction, medium properties, column packing procedures, and performance specifications.

Membranes for Tangential Flow Filtration

As with chromatography media, membranes used for tangential flow filtration should be obtained from manufacturers that adhere to the PDA's vendor certification plan (PDA 1989). Programs should be developed with the membrane supplier to ensure that all components of the filter module are consistent from lot to lot and that no substantial change in 'material or means of manufacture has occurred. The certificate of analysis for membrane modules should specify materials of construction and the results of any testing performed on the filter, including integrity testing, testing for extractables, and sterility testing (PDA 1992b).

PERFORMANCE QUALIFICATION OF DOWNSTREAM PROCESSES

Validation of downstream processes involves PQ that require rigorous testing to demonstrate the effectiveness and reproducibility of the process. The goal of PQ is to establish confidence in the performance of purification unit operations under normal as well as "worst-case" production conditions. The FDA's guideline on process validation discusses the concept of "worst-case" conditions as the extreme of normal operating conditions (FDA 1987a). The worst-case challenge in process validation should simulate those conditions that may *actually* be encountered during the production process. Test conditions for process validation should include setting processing parameters at the extreme (upper or lower processing limits) of those included within the SOPs and at values that pose the greatest probability of process failure. The purpose of worst-case testing is to demonstrate that the product and process is not adversely affected when operating parameters fluctuate within the normal, operating ranges specified in the operating procedures.

Identifying the most appropriate conditions to test to demonstrate process robustness can be a difficult task for most purification unit operations. Each operation may depend on many different variables, and these variables may not be independent. One approach to minimize the amount of testing required for validation is to apply the principles of statistical experimental design to the validation protocol (Haaland 1989; Mason et al. 1997). A specific example of the application of statistical design to the validation of process robustness is described herein on pages 139–140 (Kelley et al. 1997).

Concurrent Versus Prospective Validation

PQ is often carried out using a *concurrent validation* approach in which the process is sampled during actual production runs. Concurrent validation studies are generally carried out during the production of material for clinical trials and include expanded testing of in-process samples to gather data on the process and product. The tests are designed to document process performance and reproducibility. However, because it may be impractical and uneconomical to perform all performance studies on the actual process, a scaled-down version of the process can be used to conduct *prospective validation* studies. During these prospective studies, radiolabeled reagents, viruses, and other reagents may be employed as tracers to measure the removal of specific contaminants. When designing these experiments, the scaled-down process should be a linear extrapolation of the equipment and process parameters used in the manufacturing process.

Chromatography Column Packing

Before a chromatography column is used in production, during routine production, and after prolonged storage, it is important that the quality of the packing is checked. This may be performed by measuring the height equivalent to a theoretical plate (HETP) and asymmetry (A_s) values of the column (Berglof 1993). The test results are assessed from both the HETP value and the peak shape. The HETP value should be as low as possible, while the peak should be smooth in profile and have a high degree of symmetry with little or no tailing. It is difficult to give examples of acceptable HETP and A_s values, since they depend on the test conditions and column packing. In general, the lower the HETP value, the better; A_S values should be as close to 1.0 as possible. The choice of sample used for HETP and A_S determinations depends on the packing material. For example, size exclusion and affinity media can be tested with a 1 percent sodium chloride solution using water as the eluent. Ion exchange media may be tested with a sample of buffer that is 10 times more concentrated than the equilibration buffer. For example, for a cation exchange column equilibrated in 0.025 M sodium acetate buffer, a 0.25 M sodium acetate solution is a suitable sample. With these samples, the eluted peak is easy to measure with a flow-through conductivity meter connected to a recorder. Other innocuous reagents, such as benzyl alcohol, can be used to collect similar column separation data. Benzyl alcohol is a strong ultraviolet (UV)-absorbing compound and, thus, may provide an easy on-line measurement to assess changes in the physical state of the packed bed.

Column packing may also be assessed by analyzing the peak shape and column performance of a freshly packed column using actual production conditions (Berglof 1993; Seely et al. 1994). In this case, the elution profile of the first manufacturing run on a new column is compared to that obtained during production on a previously used column. Assuming that there are no process changes associated with the new column, the elution profile should be similar to previous profiles on the old column or to a control, thus building a historical or retrospective column record. In a virus clearance study during plasma fractionation, Berglof (1993) measured an HETP of < 0.05 and an A_S of 0.97 for a freshly packed production column. After 48 cycles, the column had an HETP of 0.029 and an A_S of 0.80, indicating that the column remained adequately stable during production.

Column Lifetime

Since production columns are frequently used for multiple batches, their cleaning, regeneration, and useful life should be validated. The most common method for validating the cleaning, regeneration, and useful life of packed columns is to produce a number of batches of product (either during development or actual production) and analyze these batches using validated analytical and biological assays. Since the lifetime of a packed column may be hundreds or even thousands of cycles, it is usually impractical to validate column lifetimes except concurrently during normal production. One approach would be to establish a set of working criteria for the evaluation of column performance and monitor this performance during production over an extended period of time. In this manner, the column's performance is evaluated with each production run against previous successful lots. With time, a database would be generated correlating column life with product quality and yield. Such a database on each column used in purifying a protein will serve as an aid to determining column life.

Specific criteria by which production columns may be evaluated before and after each use include physical, performance, and microbiological parameters. For smaller columns constructed of acrylic or glass, the column may be inspected for physical irregularities, including cracks or channels in the packed bed, discoloration of the bed, or deposits on the top of the bed. Any abnormalities observed can be recorded on a column use log and correlated to product yield and purity. Since it is common for chromatography media to become discolored with repeated use, such discoloration may be acceptable if it does not adversely affect product quality. Also, during column equilibration and production, the column back pressure can be monitored. If the column back pressure rises above a predetermined level, appropriate actions can be taken. These actions may include repacking or replacing the medium.

Seely et al. (1994) validated the reuse of ion exchange media for the purification of recombinant human interleukin-1 receptor antagonist. Their validation design included evaluation of the chromatography medium using tests similar to those used to test the incoming raw material, namely, chemical challenge with harsh cleaning solutions, evaluation of chromatography medium from production columns, small-scale cycling experiments, routine monitoring of production columns, demonstration of medium cleaning, and demonstration of the removal of leachates from the medium.

For physical and chemical evaluation of the medium, the authors measured the small ion capacity and the total protein capacity of the medium, the slope of a plot of pressure versus flow in a production column, the mean particle diameter of the medium, and total organic carbon (TOC) released after treating a sample of the medium with either 0.2 M sodium hydroxide or 1.5 M phosphoric acid. The results of these measurements for a production column packed with Q Sepharose Fast Flow (Pharmacia, Piscataway, N.J.) are shown in Table 5.5. After 143 cycles in the production column, the majority of the physical and chemical properties of the medium remain within the manufacturer's original specifications (Seely et al. 1994). A small amount of TOC remained strongly bound to the medium and was removed only by treatment with concentrations of acid or base far in excess of normal production conditions. The authors concluded that this tightly bound TOC would not be released

Table 5.5. Physical and Chemical Evaluation of Q Sepharose
Fast Flow Media Used in a Purification Process

Characteristic Measured	New Media	Production Column Measurement
Small ion capacity, mmol/g[1]	0.20	0.21
Total protein capacity, mg/g	87.0	83.0
Pressure vs. flow slope, psi min/mL	0.65	0.60
Mean particle diameter, (m	91.0	81.1
TOC released by 0.2 M NaOH, ppm	2.4	9.5
TOC released by 1.5 M phosphoric acid, ppm	1.0	44.1

[1]Manufacturer's specification for small ion capacity is 0.19–0.24 mmol/g.

Source: Seely (1994)

during normal column operations and, therefore, would not cause problems with column performance or product purity.

Column performance and lifetime can also be validated by monitoring elution profiles for consistency from run to run. Qualitative changes in a profile, such as changes in peak width or symmetry may indicate that the packing material is approaching the end of its useful life. More quantitative changes, such as peak retention times, resolution between peaks, and loss of capacity, may also be used as indicators. Conversely, process deviation, such as feed stream inconsistency, or changes in other control parameters, such as flow rate, temperature, or elution conditions, may produce altered profiles.

To ensure that changes in elution profile do not result from changes in the packed column bed, visual inspection and HETP determination can be used. Over a period of time, changes in HETP values may be correlated with column performance, and an acceptance limit set. If used, HETP measurements should be repeated when unexpected results are obtained from a column.

To validate the consistency of the elution profile for the lead column in their process, Seely et al. (1994) prepared a scaled-down column to model the performance of the production column. In their study, the small-scale column was cycled 787 times, and the elution peak volume and symmetry were recorded. Over the course of the study, the elution peak showed slight variations in these parameters. However, the product pool volume, product concentration, and yield were acceptable for each cycle measured. Additionally, there were no trends in these values, and the other parameters and elution profiles were consistent.

The cleaning and sanitization of chromatography columns may be evaluated by monitoring the bacterial endotoxin level of the column eluate prior to sample loading. For example, if all process buffers have a bacterial endotoxin level less than some predetermined value, such as 0.25 EU/mL, then the column would be acceptable for use if, after cleaning, sanitization, and reequilibration, the column eluant has an endotoxin level less than or equal to that of the process buffers. Similarly, the endotoxin level of the purified product recovered from the column should consistently be within a predetermined range. If the endotoxin level of the column eluant after cleaning or the purified product recovered rises above a predetermined level, it may be necessary to replace the medium.

Additionally, periodic blank chromatography runs should be performed in which the column is eluted with production buffers, and samples collected and tested to determine the extent of product and/or impurity carryover from run to run on the column. In the study of Seely et al. (1994), a blank run was performed after 142 production cycles on a Q Sepharose Fast Flow column. Samples were collected throughout the region where the target protein peak would normally elute. These samples were then analyzed by high performance liquid chromatography (HPLC), enzyme-linked immunosorbent assay (ELISA), and sodium dodecyl sulfate–polyacrylamide gel electrophoresis (SDS–PAGE). These assays indicated the presence of the target protein in the samples. However, since the amount of material represented only 0.3 percent of the amount typically collected during a production run, the authors felt that continued use of the column in production was acceptable.

A final issue that should be evaluated during column lifetime studies is the leaching of material from the chromatography medium. Even with the availability of substantial data supporting the level of leaching from various chromatography media (Johansson 1992), it is still important for users to demonstrate the stability of the media in their particular applications. For chromatography media, such as Q Sepharose Fast Flow, with polysaccharide backbones, a modified anthrone assay can be used to analyze for leachates (Scott and Melvin 1953). Seely et al. (1994) found that even after exposure to 0.5 M sodium hydroxide at 60°C for several months, only 2–3 ppm of carbohydrate was released from Q Sepharose Fast Flow medium. When this medium was packed into chromatography columns, the very low levels of carbohydrate were removed after washing the column with three column volumes of buffer.

For affinity media prepared by covalently coupling a biospecific ligand to a support medium, it is important to demonstrate that the coupling chemistry is stable enough to prevent ligand (protein) leaching from the support during purification processes. This ligand leakage may result from either chemical cleavage of the covalent bond or physical sloughing of medium fragments containing ligand. Ligand leakage can change with time, and leakage may be higher when the medium is new. In either case, the released ligand becomes a contaminant in the product. Since these ligands may be recombinant proteins or monoclonal antibodies, it is important that the ligand be of high purity, well characterized, and produced by validated processes before coupling to the chromatographic support.

In order to detect ligand leakage, appropriate assays with high sensitivity may need to be developed to measure the level of ligand in the product. In addition to monitoring ligand leakage in the actual production process stream, radiolabeled ligand can be coupled to a support on a small scale, and leakage rates can be monitored under continuous exposure to binding and elution conditions.

Ligand leakage in the product can be minimized by choosing the appropriate support and coupling chemistry and conditioning the column by repeated washing with binding and elution buffers prior to production. Also, inclusion of additional processing steps after the affinity step that specifically removes leaked ligand from the product may provide assurance that the final product will be free of the affinity ligand. If such affinity processing steps are included, clearance studies demonstrating the ability of the process to remove leaked ligand from the product should be performed.

Process Robustness

The most important criterion for validating a chromatography step is demonstrating that when the column is operated in a specified manner, the process step yields a product of consistent quality that conforms to specifications. Validation should demonstrate that the process is robust and will not fail when carried out within the normal operational range of critical process parameters. Validation of some of the operational ranges of process variables may be conducted either at scale or in scaled-down experiments. In a process robustness study, all process variables, such as buffer pH and ionic strength, gradient shape, amount of material applied per unit volume of packing material, temperature, flow rate, and system back pressure, should be considered. These studies may be conducted by changing one parameter at a time with all other parameters fixed or by using a multifactorial design in which many parameters are changed in a systematic way.

Kelley et al. (1997) used a multifactorial design to conduct a process robustness study of the cation exchange purification of recombinant human IL-11 (rhIL-11) produced in *Escherichia coli.* In this study, process goals for product yield and purity were established, and five major process variables were identified that may affect these goals. These process variables included the mass of rhIL-11 loaded onto the column (load mass), the purity level of the rhIL-11 loaded onto the column (load purity), the conductivity of the column load solution (load conductivity), and the pH and conductivity of the elution and wash buffers (elution pH and conductivity and wash pH and conductivity, respectively). To determine the effect of these variables on the ion exchange step, a 16-run resolution V study was performed examining various combinations of the process variables (see Table 5.6).

For the validation study, eight unique column load solutions were prepared based on different combinations of three load variables (load mass, load purity, and load conductivity). Each of these eight column load conditions was then tested using one of two combinations of column wash and elution buffers (elution pH and conductivity and wash pH and conductivity, respectively). The specific combinations tested are listed in Table 5.6, along with the purity and yield of rhIL-11 recovered in each case. Using an analysis of variance (ANOVA) to determine statistical significance of the variables and to determine worst-case predictions of product yield and purity, the authors demonstrated that process variation within the ranges tested produced a product with acceptable yield and appropriate purity. Because variables tested in this study represented the most probable causes of process variation for this chromatography step, the study provided a high degree of assurance that the step will perform adequately despite normal variations in feed stream, buffer composition, and column operation (Kelley et al. 1997).

CLEARANCE STUDIES

Purification processes typically contain a number of operations designed to inactivate or remove viral, nucleic acid, immunogenic, and

Table 5.6. Process Robustness[1]

Run #	Load Mass[2]	Load Purity[3]	Load Conductivity[4]	Elution pH & Conductivity[5]	Wash pH & Conductivity[6]	Product Recovery, %	Product Purity, %
1	-1	-1	-1	1	-1	112.6	96.3
2	-1	-1	-1	-1	1	90.7	96.9
3	-1	-1	1	1	1	104.9	97.1
4	-1	-1	1	-1	-1	72.8	97.3
5	-1	1	-1	1	1	99.6	96.1
6	-1	1	-1	-1	-1	84.2	97.1
7	-1	1	1	1	-1	98.4	97.3
8	-1	1	1	-1	1	104.2	97.8
9	1	-1	-1	1	1	104.3	91.8
10	1	-1	-1	-1	-1	79.0	94.9
11	1	-1	1	1	-1	94.8	96.6
12	1	-1	1	-1	1	93.7	96.0
13	1	1	-1	1	-1	95.5	94.4
14	1	1	-1	-1	1	88.5	93.9
15	1	1	1	1	1	78.7	96.5
16	1	1	1	-1	-1	58.7	98.4

[1] In each experiment, the indicated process parameter was adjusted to either the maximum (1) or minimum (-1) allowable value as specified in the production documentation for this process step.

[2] Lower limit (-1) = 2.4 g rhIL-11/L chromatography media; Upper limit (1) = 15.5 g rhIL-11/L chromatography media

[3] Lower limit (-1) = 63; Upper limit (1) = 75

[4] Lower limit (-1) = 2.5; Upper limit (1) = 4.2

[5] Lower limit (-1) = 9.4 (pH) and 8.6 (conductivity); Upper limit (1) = 9.6 (pH) and 14.4 (conductivity)

[6] Lower limit (-1) = 9.4 (pH) and 2.5 (conductivity); Upper limit (1) = 9.6 (pH) and 4.2 (conductivity)

Source: Kelley et al. (1997)

pyrogenic contaminants without affecting the biological activity of the desired product. In addition to those contaminants that may have been present in the initial bioreactor harvest, other contaminants, such as reagents used during purification or ligands that may have leached from chromatography medium used in the purification of the product, must also be removed from the product during downstream processing. Successful measurement of elimination or inactivation of contaminants can be determined by specific assays such as radioimmunoassays (RIAs), enzyme immunoassays, and protein blotting directed toward the contaminants. Therefore, both endpoint testing and clearance studies demonstrating the removal of specific contaminants should be included as part of process validation to provide assurance that the process will effectively eliminate or inactivate specified contaminants from the product.

If noxious or infectious agents are to be used during clearance studies, it is unwise to allow these materials to be introduced into a production facility or equipment. These agents may contaminate clinical or commercial product or place manufacturing workers at unnecessary risk. Instead, these studies should be conducted on smaller-scale equipment, where the process is accurately reproduced to ensure that the data generated can be extrapolated to production-scale equipment (PDA 1992a). Typically, clearance studies are performed on chromatography operations, as these steps are generally responsible for contaminant removal. To accomplish the scale-down of a chromatography operation in an effective manner, the column medium under test should be of the same type and, preferably, the same production lot as that used in the process-scale column. Furthermore, all significant process parameters should be maintained at constant levels. In its guideline for viral safety, the International Conference on Harmonisation (ICH) states that "the level of purification of the scaled-down version should represent as closely as possible the production procedure." For chromatographic equipment, column bed height; linear flow rate; flow rate to bed volume ratio (i.e., contact time); buffer and gel types; pH; temperature; and the concentration of protein, salt, and product should all be shown to be representative of commercial-scale manufacturing (ICH 1995).

The flow rate used in a clearance study should be scaled down by the ratio between the cross-sectional area of the production column and the scaled-down column so that the linear velocity remains constant. The column bed height should remain the same as that used in production so that the contact time of the feed solution with the medium is not altered. For adsorption separations, gradient slope and volume should be scaled down by the ratio of the total volume of the production column to the volume of the scaled-down column. The ratio of product loaded to column volume should be kept constant, and the product should be present during the tests at the same relative concentrations present during the actual manufacturing process. Finally, to be valid, the yield and purity of the product recovered from the scaled-down column should be consistent with that of the production column. The extent to which a given column-based separation is scaled down for validation will depend on the actual production scale and the smallest scale that can reliably reproduce the production process.

In clearance studies, a particular contaminant is added to the input feed stream on a small scale, and the recovery of the contaminant is measured at each stage of the process step, such as the column flow through, product pool, and regeneration fractions using scaled-down columns (PDA 1992a). The addition of the contaminant should be kept to a minimum so as to not significantly change the concentration of the feed stream or alter the behavior of product recovery. Measurements for mass balance calculations should be performed on column flow through (nonbinding materials), eluted fractions, regeneration, and cleaning steps. Mass balance during the regeneration and cleaning steps is critical in assessing whether or not the column packing material can be reused.

A clearance factor can be calculated as shown in Equation 1 by dividing the number of units introduced by the number of units recovered in the product after that step.

$$CF_i = I \div O \tag{1}$$

In Equation 1, CF_i is the clearance factor for the ith step in the process, I represents the number of units introduced at the start of the process step, and O represents the number of units recovered after the process step.

Each step in a purification process should be challenged separately so that the clearance of a particular contaminant by each step of the process can then be calculated. In general, the overall clearance factor for a manufacturing process (CF_t) is the product of the clearance factors for each step:

$$CF_t = (CF_1 \times CF_2 \times CF_3 \dots CF_n) \tag{2}$$

When radiolabeled tracers are used in clearance studies, the interpretation of clearance factors may be more complex. If the tracer is a homogeneous species, or if the tracer behaves as if it were a homogeneous species in the process under study, then the clearance of the tracer in each step is independent of the sequence in which the steps are performed. In this case, the clearance factors measured at each step in the process may be multiplied together, with the resulting product representing the clearance factor for the entire process.

If the tracer used is not homogeneous, then the interpretation of clearance factors may be more difficult. [^{32}P]DNA, which is commonly used in clearance studies, is an example of a heterogeneous tracer. It is a chemically diverse population (i.e., the population consists of molecules of different nucleotide sequences of various lengths, with a distribution of molecular weights). Some of the separation methods used in the purification of recombinant proteins are insensitive to either the nucleotide sequence or the molecular weight distribution. For these processes, the assumption that [^{32}P]DNA behaves as a single homogeneous species is a valid one, and the clearance factors obtained at each step may be multiplied together to give an overall clearance factor for the process.

Radiolabeled host cell protein is another example of a chemically diverse population of molecules that may behave as a heterogeneous population in protein purification processes. The population may consist of several hundred labeled proteins that are heterogeneous with respect to charge, hydrophobicity, thiol content, and molecular weight. Therefore, in these cases, the practice of multiplying clearance factors obtained from each individual step to obtain an overall clearance factor may have no practical significance, and care should be taken in interpreting the results of such experiments (Hageman 1991).

The principle contaminants that may require clearance studies are pyrogens, media components, host cell proteins, nucleic acids, viruses, and materials leached from bioaffinity media.

Nucleic Acids

The concern of potential biohazards from the presence of nucleic acids in parenteral protein preparations led to the introduction of regulatory guidelines that limit the exposure of patients to DNA. Original guidelines from the World Health Organization (WHO) and the FDA placed these limits at 10–100 pg per dose per day (FDA 1985, 1987c; WHO 1987). However, with experience accumulated to date, the perceived risk from DNA contamination is now not very large, and these initial guidelines have been somewhat relaxed (FDA 1993, 1997a). Despite this relaxation in regulatory guidelines, there is still a requirement to demonstrate the removal of nucleic acids from biopharmaceutical products, especially in protein preparations where large doses of protein will be administered to patients. Typically, validation of DNA removal includes monitoring the elimination of source DNA in the process stream at each key purification step and confirmation of removal in the final product using a sensitive residual DNA assay.

Several analytical techniques are available for detecting DNA in in-process samples and final product. Sequence specific DNA can be detected by hybridization using specific probes to measure product-related DNA (Eaton 1989; Ostrove et al. 1992; Per et al. 1989) or the Threshold System (Molecular Devices Corporation, Sunnyvale, Calif.) to measure total DNA using a molecular sensor-based assay (Kung et al. 1990), but they are not sufficiently sensitive to detect trace levels of DNA in a single dose. Because of higher specificity, the techniques for detecting specific DNA sequences often have a better signal-to-noise ratio and, therefore, are able to detect smaller quantities of DNA than the total DNA method (Riggin and Davis 1996). Nevertheless, the total DNA detection method can be very useful in validation studies, especially where many samples must be analyzed to demonstrate the removal of DNA effectively.

To determine the removal of DNA by downstream processing, clearance studies using scaled-down production equipment and radiolabeled DNA are typically performed. In these studies, production source DNA, extracted from the host strain or cell line, is radiolabeled by nick translation (Rigby et al. 1977). After separating free radioactivity from the labeled DNA, the labeled DNA is spiked into the feed stream to be tested, and the spiked sample is subjected to the scaled-down purification operation.

Because of the highly acidic nature of nucleic acids, they are most readily removed during downstream processing by ion exchange chromatography. For clearance studies, the scaled-down column is run under conditions that closely mimic the production column, and fractions are collected during the chromatographic run. Radioactivity is measured in every fraction, and the elution profile for nucleic acids is compared to that of the product of interest. When performing clearance studies such as these, it is important that the labeled DNA be essentially free of low molecular weight radioactivity that will bias the results. The size of the labeled DNA should be similar to that found in the initial crude extracts or conditioned medium, since the ability of different chromatographic steps to clear DNA can be a function of the molecular weight of the nucleic acid. Clearance factors are determined by calculating the ratio of radioactivity loaded on the column to the radioactivity contained in the product fraction(s).

An example of DNA removal from a recombinant protein is shown in Table 5.7 (Berthold and Walter 1994). In this particular validation study, 21 consecutive purification cycles were performed using 3 different anion exchange chromatography media. Radiolabeled DNA was spiked into the column load before each cycle and after every 5 cycles, and the clearance factor for the removal of all DNA and DNA of greater than 50 base pairs was determined. For each chromatography medium, the clearance factor was consistent throughout the validation study. The average clearance factor obtained for two Sepharose Fast Flow anion exchangers (Q Sepharose and DEAE Sepharose) was approximately 1.5 million. For DE-52 Cellulose (Whatman, Clifton, N.J.), the clearance factor was approximately half that of the Sepharose exchangers or 0.7 million (Berthold and Walter 1994). Each of these validation studies demonstrated that a final concentration of DNA of less than 100 pg per dose of protein could be reproducibly achieved.

Another example of DNA removal determined using radiolabeled DNA in a scaled-down clearance study is shown in Table 5.8 (Banks et al. 1989a). In this study, a clearance factor for the removal of DNA from a murine monoclonal antibody produced in ascites fluid was determined for each chromatographic step in the purification process. This process of 3 chromatography steps (Protein A affinity chromatography, S Sepharose Fast Flow, Q Sepharose Fast Flow) has an overall clearance of approximately 5 logs. This number is based on the removal of radiolabeled DNA spiked into individual column loads and is in close agreement with the analysis of individual in-process samples from production runs using a hybridization assay (Banks et al. 1989a). Since crude ascitic fluid for this monoclonal antibody typically contains between 100 and 1,000 ng DNA/mL and an antibody concentration of 2 mg/mL, a 5 log clearance would reduce DNA levels in the purified antibody to 0.5–5 pg/mg. Assuming a therapeutic dose of 10 mg of the monoclonal antibody, the final DNA level in the product would be 5–50 pg/dose, well below the acceptable level of 100 pg/dose.

Host Cell Proteins

The clearance of host cell proteins may be measured by using a direct immunoassay of these proteins in the actual process stream. Production source proteins are prepared from the host organism or cell that either contains a plasmid constructed to have all of the DNA sequences except those for the gene encoding the protein product or the parent myeloma cell of a hybridoma. These proteins are isolated, and polyclonal antibodies are prepared against them. Alternatively,

Table 5.7. Validation of DNA Removal by Anion Exchange Chromatography[1]

	Elution Factor[2]	Clearance Factor[3]
Q Sepharose Fast Flow		
Cycle #1	35,700	2,360,000
Cycles #2–5	—	—
Cycle #6	14,500	1,500,000
Cycles #7–10	—	—
Cycle #11	18,500	1,320,000
Cycles #12–15	—	—
Cycle #16	24,500	1,500,000
Cycles #17–20	—	—
Cycle #21	28,600	1,050,000
Average	24,360	1,546,000
DEAE Sepharose Fast Flow		
Cycle #1	31,700	712,000
Cycles #2–4	—	—
Cycle #5	20,400	750,000
Cycles #6–8	—	—
Cycle #9	23,200	2,600,000
Cycles #10–18	—	—
Cycle #19	18,100	1,300,000
Average	23,350	1,340,000
DE-52 Cellulose		
Average of 7 runs	15,000	700,000

[1]Column dimensions = 1 × 20 cm (15.7 mL); protein load = 10 mg/mL chromatography medium; DNA load = 7 µg/mL chromatography medium; linear flow rate = 150 cm/h

[2]The elution factor represents the total radioactivity that coelutes with the protein and does not discriminate between DNA molecules of different chain lengths.

[3]The clearance factor represents radioactivity precipitable with 15% trichloroacetic acid (TCA). Only DNA molecules greater than 50 base pairs are precipitable with TCA.

Source: Berthold and Walter (1994)

Table 5.8. DNA Clearance from a Monoclonal Antibody by Chromatography

Process Step	DNA Level in Column Load[1]	Estimated Clearance Factor[2]	Clearance Factor from Validation Study[3]
Protein A Affinity	413 ng/mg	3.6 log	2.0 log
S Sepharose Fast Flow	109 ng/mg	1.2 log	2.0 log
Q Sepharose Fast Flow	7.75 pg/mg	0	1.0 log
Bulk Antibody	22 pg/mg	—	—
Total process clearance		4.8 log	5.0 log

[1]Samples of each column load solution from three different production runs were analyzed by a hybridization assay.

[2]Clearance factors were estimated by comparing the estimate of DNA content in each load solution to the DNA content of the load for the next purification step.

[3]The clearance factor from the validation study was determined by comparing the amount of radiolabeled DNA that eluted from the column with the total amount of radiolabeled DNA applied to the column.

Source: Banks et al. (1989a)

clearance of host cell proteins can be determined by radiolabeling these production source proteins, and then adding or "spiking" them into the appropriate crude feed stream or intermediate stream to the column under scaled-down test conditions. The total radioactivity of all fractions collected is determined and compared to the total loaded radioactivity for mass balance determinations. Fractions containing product are then pooled, and the clearance factor is calculated. The clearance factor is the ratio of total loaded radioactivity to radioactivity contained in the product fraction. The overall clearance factor for the purification process will vary according to specificity, selectivity, and the number of process steps.

Pyrogens

Because of their high molecular weight and highly negative charge, pyrogens are commonly removed from proteins by either ion exchange chromatography, gel filtration, or ultrafiltration (Minobe et al. 1983; Nolan et al. 1975; Novitsky and Gould 1985; Schindler and Dinarello 1990). Among these methods, ion exchange chromatography is generally the most useful in reducing bacterial endotoxin levels, provided that the selectivity of the medium is such that copurification of the bacterial endotoxin and product is avoided. The Limulus amebocyte lysate (LAL) assay for gram-negative bacterial endotoxins is sensitive enough to detect concentrations at least an order of magnitude below levels that will produce a pyrogenic reaction in the rabbit pyrogen test. The possibility of inhibition or enhancement of the LAL assay by the protein product, however, must be ascertained through validation of the LAL test (FDA 1987b).

Since a sensitive assay is available to detect the presence of pyrogens in in-process samples and final drug preparations, clearance studies demonstrating the removal of pyrogens may not be necessary. Good process control and hygiene (i.e., LAL testing of all raw materials, microfiltration or ultrafiltration of process buffers, and cleaning and sanitation of columns after each use) will minimize the potential for endotoxin contamination and, hence, the need for clearance studies. If the performance of clearance studies is desired, they may be carried out following the guidelines described above.

Viruses

When mammalian cells are used as substrates to produce a protein product, there is concern that the cell lines may harbor viruses (Brown 1988; Lubinecki et al. 1990). Endogenous retroviruses are widespread in animal populations and have been described in species as diverse as reptiles, birds, and many mammals. For example, murine hybridomas used in the production of monoclonal antibodies are known to express retroviruses that may have the potential to transform cells. Other rodent cell lines such as CHO and baby hamster kidney (BHK) may also contain these endogenous retroviruses. In the absence of a specifically identified viral contaminant in the product cell line, the potential presence of retroviral particles is of greatest concern. In addition, concerns regarding bovine viruses and prions are also increasing among regulatory agencies (WHO 1997; FR 1993; FDA 1996d).

The most appropriate way to ensure that viruses do not copurify with product is to test and select production cells and media components that are free from known adventitious viral contamination. Since most cell lines currently used in production are derived from sources that cannot be certified as free of endogenous viruses, and since adventitious agents may enter the production process and propagate in cells, viral clearance studies for products derived from cell cultures are essential. Viral clearance is most readily measured by small-scale spiking experiments. It should include both virus removal and inactivation and clearance factors several logs greater than the theoretical titer of infectious virus per dose of product (Brown and Lubinecki 1996). A theoretical worst-case titer may be estimated from electron microscope pictures of the cell culture fluids from which the product is purified. This information, combined with the process yields and the expected dose size, is used to compute the number of viruses that would be carried into the dosage unit if there were no clearance by the purification process (Lubinecki et al. 1990).

In addition to characterizing the viruses contained in the cell line, it is important to demonstrate that the purification process can remove and/or inactivate those viruses that may be indigenous to the cell line but remain undetected. Therefore, it is desirable to perform spiking experiments with viruses that can be cultivated to a high titer, that have well-established detection assays, and that do not present health hazards.

For proteins produced by recombinant DNA technology or naturally by human cell lines, validation of virus removal or inactivation should include a collection of model viruses possessing a range of biophysical and structural features. Table 5.9 lists several viruses that have been used in virus validation studies (PDA 1992a). The viruses used should include enveloped and nonenveloped DNA and ribonucleic acid (RNA) viruses that have different diameters and geometries.

Table 5.9. Examples of Viruses That Have Been Used in Virus Validation Studies

Virus	Family	Natural Host	Genome	Enveloped	Size	Shape	Resistance to Physicochemical Reagents
Poliovirus, Sabin type 1	Picorna	Man	RNA	No	25–30 nm	Icosahedral	Medium
Reovirus 3	Reo	Various	RNA	No	60–80 nm	Spherical	High
SV-40	Papova	Monkey	DNA	No	45 nm	Icosahedral	High
Murine leukemia virus	Retro	Mouse	RNA	Yes	80–110 nm	Spherical	Low
HIV	Retro	Man	RNA	Yes	80–100 nm	Spherical	Low
Vesicular stomatitis virus	Rhabdo	Bovine	RNA	Yes	80–90 nm	Bullet Shaped	Low
Parainfluenza virus	Paramyxo	Various	RNA	Yes	150–300 nm	Pleo-Spher	Low
Pseudorabies virus	Herpes	Swine	DNA	Yes	120–200 nm	Spherical	Medium

Source: PDA (1992a)

DNA viruses such as herpes simplex 1 (enveloped) and SV-40 (nonenveloped) and RNA viruses such as Sabin type I polio (nonenveloped) and influenza type A (enveloped) represent typical challenge viruses. When rodent cells such as CHO, BHK, C127, and murine hybridomas are used for production, then Moloney murine leukemia virus may be used as a model retrovirus. When choosing an appropriate challenge virus, preference should be given to those viruses that display a significant resistance to physical and/or chemical agents.

Clearance studies similar to those described above for host cell proteins and DNA may be performed by spiking model viruses into the production stream and measuring their removal on scaled-down columns. The clearance of virus particles may also be measured using radiolabeled virus. Radiolabeled virus can be prepared in a similar manner to the preparation of labeled host cells, using [^{3}H]-, [^{14}C]-, or [^{35}S]-labeled amino acids. As mentioned above, care should be taken to prepare labeled virus that is free from molecular weight–labeled contaminants. Each stage of the purification process should be individually assessed for its ability to remove or inactivate virus. The overall clearance factor can be determined from individual clearance factors. Care should be taken in calculating the overall clearance factor, however. The assumption that clearance factors of different steps may be multiplied to give the overall clearance factor may not always be valid. Clearance factors may only be additive, for example, if the mechanism of virus removal in two different steps is the same.

Since membrane-enveloped virus may shed surface proteins, the assay for virus particles should include steps to distinguish viral particles from shed proteins. Alternatively, since retroviruses contain a specific enzyme marker, reverse transcriptase, it may be possible to demonstrate their clearance through the use of an enzymatic or immunologic assay for reverse transcriptase. However, the reverse transcriptase assay is inaccurate at low concentrations; in crude samples, care should be taken to avoid interference from cellular DNA polymerases.

Grun et al. (1992) provided a summary of virus removal by a variety of purification methods. Average log clearance factors ranging from approximately 1.3 to 5.1 were noted for a variety of chromatography types (Table 5.10). However, within each type of chromatography, the range of viral clearance varied widely and depended on the specific virus tested and the exact purification process used. This wide range of virus removal was also noted in a more recent review (Darling and Shapiro 1996).

In addition to demonstrating the removal of viral particles, virus inactivation should also be measured. Retroviruses are labile

Table 5.10. Log$_{10}$ Virus Titer Reduction by Column Chromatography

Chromatography Mode	E-MuLV[1]			X-MuLV[2]			HSV[3]			Polio		
	X[4]	SD[5]	Range	X	SD	Range	X	SD	Range	X	SD	Range
Affinity/adsorbent	2.5	1.3	1.1–4.2	2.2	0.90	1.2–3.6	2.2	0.3	2.0–2.3	1.6	2.2	0–3.1
Anion exchange	3.7	1.5	1.8 - 5.7	2.0	1.0	1.0–3.1	5.1	0.0	5.1[6]	2.6	0.0	2.6[6]
Cation exchange	2.8	1.4	1.2–4.8	2.0	0.4	1.6–2.4	—	—	—	—	—	—
Gel filtration	1.3	1.5	0.2–4.0	2.4	0.7	1.9–2.4	3.4	2.4	0.6–5.2	0.8	0.8	0 - 1.6
Mixed-mode exchange	—	—	—	2.9	0.8	2.3–3.5	3.0	0.2	2.9–3.1	2.7	1.1	1.9–3.5
Hydrophobic interaction	3.2	0.4	2.8–3.8	4.1	0.0	4.1[6]	4.0	0.3	3.8–4.2	2.8	0.8	2.2–3.4
Hydroxyapatite	2.0	0.0	2.0[6]	0.6	0.0	0.6[6]	—	—	—	—	—	—

[1]Ecotropic murine leukemia virus
[2]Xenotropic murine leukemia virus
[3]Herpes virus
[4]Mean
[5]Standard Deviation
[6]Individual values were not significantly different

Source: Grun et al. (1992)

species; a well-designed process may include steps that can be validated as virus inactivation steps. A column-based separation may provide viral inactivation as well as removal, especially if nonneutral pH, denaturing reagents, or organic solvents (as used for HPLC) are used. To demonstrate virus inactivation, the virus may be spiked into a process solution and incubated under time and temperature conditions that model the normal production process. When conducting these inactivation studies, it is desirable to determine the kinetics of inactivation as well as the extent of inactivation, because virus inactivation has been demonstrated in some cases to be a complex reaction with a "fast phase" and a "slow phase" (EC 1991). The inactivation study should be performed in such a way that samples are taken at different times in order to construct an inactivation curve. To do these studies, a high-titer virus stock is needed, as well as the appropriate infectivity assay.

A specific example of a viral clearance study is described by Berthold and Walter (1994) and is detailed in Table 5.11. In this study, clearance studies were performed using four different viruses in each of 3 chromatography steps, an ultrafiltration step, and a specific viral inactivation step. Each process step was evaluated individually, with greater than 6 logs of each virus spiked into the process feed stream prior to performing the separation or inactivation process. Consistent with the results of Grun et al. (1992) and Darling and Shapiro (1996), the anion exchange chromatography steps each resulted in 4–6 logs of virus clearance. Very little clearance (< 1 log) was seen for the cation exchange chromatography step, and approximately 3 log clearance was seen in the ultrafiltration step using a 300,000 NMWL (nominal molecular weight limit) membrane. For the specific viral inactivation step, a clearance of 3–5 logs was seen, depending on the specific virus tested.

A virus clearance study of a monoclonal antibody purification process is outlined in Table 5.12 (Banks et al. 1989b). Using either xenotropic or ecotropic viruses, the overall clearance factor for the process steps tested was approximately 9 logs. Because antibody was eluted from the Protein A affinity chromatography column using a pH 3.0 buffer, the virus clearance from this column is a combination of removal by the affinity column and inactivation by the low pH buffer used for product elution. In an attempt to differentiate between virus removal and inactivation, the clearance study was repeated using radiolabeled virus particles prepared by growing the appropriate hybridoma cells in media containing 5-[^{3}U]-uridine and [^{14}C (U)]-L-amino acids. The radiolabeled viral particles were purified by equilibrium density centrifugation and then spiked into the

Table 5.11. Results of a Virus Clearance Study

Process Step	Spike[1]	Clearance Factors			
		Xenotropic Retrovirus	PI 3	Reo	SV-40
Anion Exchange	6.6	4.3	ND[2]	6.1	6.0
Cation Exchange	6.6	< 1	ND	< 1	< 1
300 kd Ultrafiltration	6.4	3.6	ND	3.0	< 1
Gel Filtration	6.5	1.8	ND	2.2	1.8
Virus Inactivation	6.5	4.6	5.4	< 1	3.1
Cumulative Reduction		14.3	5.4	11.3	10.9

[1]Virus titers and clearance factors are given in log 10 PFU/mL.

[2]Not tested

Source: Berthold and Walter (1994)

column load. When the affinity chromatography column was run using this spiked material, the bulk of the radioactivity flowed through the column during the sample load and wash of the column. Only trace amounts of radioactivity were found in the column pool, thus indicating that the bulk of the virus was separated from the antibody before it was eluted from the column.

VALIDATION OF TANGENTIAL FLOW FILTRATION

Because of its wide variety of applications (Gabler 1986; Lonsdale 1982; McGregor 1986), it is difficult to provide simple guidelines for the validation of tangential flow filtration (TFF). Validation of TFF will generally include a verification that the molecular weight cut-off for ultrafiltration membranes or pore rating for microfiltration membranes is appropriate for the intended use, along with a verification that the membrane used meets the specifications for the process step. In addition, critical process parameters for TFF, such as

Table 5.12. Results of a Virus Clearance Study for a Murine Monoclonal Antibody

	Clearance Factors	
Process Step	**Murine Ecotropic Retrovirus**	**Murine Xenotropic Retrovirus**
Protein A affinity chromatography	>3.52	>3.10
S Sepharose Fast Flow	2.69	2.43
Q Sepharose Fast Flow	>3.32	>3.11
Cumulative reduction	>9.53	>8.64

[1]Virus titers and clearance factors are given in log 10 PFU/mL.

Source: Banks et al. (1989b)

filtrate and retentate pressures, fluxes, and temperatures, should be verified during validation (PDA 1992b). The following examples of the validation of tangential flow operations should serve as a general guideline for specific applications.

Protein Concentration/Diafiltration

When a TFF system is used for protein concentration or buffer exchange, the primary process parameters to measure include the operating ranges of pressure, flow, and temperature. For diafiltration, the removal of low molecular weight contaminants should also be monitored.

During validation, the operating ranges under which the product does not undergo adverse changes, such as aggregation, denaturation, or loss of activity, should be defined. It may also be useful to investigate the effect of changes in flux on product recovery and activity (PDA 1992b).

Cell Debris Removal

When a TFF system is used for the removal of cell debris, or other liquid/solid separations, the product is generally found in the permeate. If this TFF operation is only a clarification step, process qualification should address issues of permeate clarity, product quality, and yield. In this case, validation needs only demonstrate that the process stream is clarified to an acceptable level and contains the product in acceptable and consistent yields (PDA 1992b).

If the TFF process is actually a purification step, then validation must also establish that the degree of purification is acceptable and consistent. In these operations, membrane polarization and fouling may occur under certain operating conditions, which may limit the effectiveness of the purification. Consequently, the process should be tested at the extremes of the manufacturing operating ranges for tangential flow rates, pressures, and flux to verify that such fouling is not occurring and does not adversely affect product quality (PDA 1992b). For the removal of specific contaminants, clearance studies such as those described above should be conducted.

Cell Harvesting

Cell harvesting is a straightforward application of TFF (McGregor 1986) and should be validated similarly to cell debris applications (PDA 1992b). For the harvesting of cells where the product is retained intracellularly, validation should also include a demonstration that the host organism is adequately contained during the operation. If the permeate during a cell harvesting operation is considered nonsterile, then validation must show that the system hardware (tanks, piping, valves, etc.) and operating procedures do not allow the escape of the recombinant organism. If the permeate is treated as a solution free from the host organism, validation of the TFF operation should include a demonstration that the TFF system is capable of consistently removing the entire bacterial load from the harvest feed stream (PDA 1992b). Furthermore, in-process testing should include integrity tests of the filter membranes before and after each use (Gabler 1986).

Process validation of a cell separation operation where the product is secreted into the medium should also ensure that the process delivers consistent product quality and yield. In any cell removal system, product quality could be adversely affected by cell lysis, which could lead to proteolytic degradation of the product and/or decreased product purity, by increases in harvest time, by trace impurities, and filtration membrane integrity (PDA 1992b). For these cell separation operations, the integrity of the filter membranes should be tested before and after each use, and the level of contaminants should be shown to be consistent from batch to batch (Gabler 1986; Werner et al. 1988).

SUMMARY

The validation of biopharmaceutical manufacturing processes is necessary to ensure the quality and safety of biologic products. While this chapter has emphasized the validation of manufacturing processes for recombinant proteins, the principles and approaches presented are relevant to all biopharmaceutical manufacturing processes. As with traditional pharmaceutical manufacturing processes, process validation for biologic products includes the qualification of raw materials and equipment and assays followed by performance testing of the fermentation and purification process. Combined with in-process control and QC of the final product, process validation ensures that a uniform product is produced consistently from batch to batch.

Validation should be designed into a manufacturing process from the outset. During process development, techniques should be selected that can optimize product formation and remove impurities and contaminants. The equipment and raw materials used in these operations should be carefully selected so that the process will perform consistently and reproducibly. Once the process is optimized and production begins, validation of critical process parameters will ensure the reliable and reproducible production of the biologic product.

REFERENCES

Aiello, L., R. Guilfoyle, K. Huebner, and R. Weinmann. 1979. Adenovirus 5 DNA sequences present and RNA sequences transcribed in transformed human embryo kidney cells (HEK-Ad-5 or 293). *Virology* 94:460–469.

Asenjo, J. A., and J. C. Merchuk, eds. 1995. *Bioreactor system design.* New York: Marcel Dekker, Inc.

Banks, J. F., K. Gikonoyo, R. T. Kawahata, and H. L. Levine. 1989. Unpublished data.

Banks, J. F., J. Thrift, F. J. Castillo, J. M. Sadowski, R. T. Kawahata, and H. L. Levine. 1989b. Unpublished data.

Bebbington, C. R., and K. Lambert. 1994. Genetic stability and product consistency of rDNA–derived biologicals from mammalian cells. *Develop. Biol. Stand.* 83:183–184.

Berglof, J. H. 1993. Validation aspects relating to the use of chromatographic media. *Colloque INSERM* 227:31–36.

Berthold, W., and J. Walter. 1994. Protein purification: Aspects of processes for pharmaceutical products. *Biologicals* 22:135–150.

Blau, H., and P. Khavari. 1997. Gene therapy: Progress, problems, prospects. *Nature Med.* 3:612–613.

Bolin, S. R., J. F. Ridpath, J. Black, M. Macy., and R. Robin. 1994. Survey of cell lines in the American Type Culture Collection for bovine viral diarrhea virus. *J. Virol. Meth.* 48:211–221.

Bremmer, R. E. 1986. Calibration and certification. In *Validation of aseptic pharmaceutical processes,* edited by F. J. Carleton and J. P. Agalloco. New York: Marcel Dekker, Inc., pp. 47–91.

Brown, F., and Lubinecki, A., eds. 1996. *Viral safety and evaluation of viral clearance from biopharmaceutical products.* New York: Karger.

Brown, J. 1988. Safety aspects of licensing biotechnology products intended for use in man. In *Proceedings of BioSymposium,* 15–20 October, in Tokyo.

Brunkow, R., D. Delucia, S. Haft, J. Hyde, J. Lindsay, J. McEntire, R. Murphy, J. Myers, K. Nichols, B. Terranova, J. Voss, and E. White. 1996. *Cleaning and cleaning validation: A biotechnology perspective.* Bethesda, Md., USA: Parenteral Drug Association.

Calcott, P. H. 1996. Cryopreservation of microorganisms. *CRC Crit. Rev. Biotechnol.* 4:279.

Castillo, F. J. 1995. Organism selection. In *Bioreactor system design,* edited by J. A. Asenjo and J. C. Merchuk. New York: Marcel Dekker, Inc., pp. 13–45.

Castillo, F. J., L. J. Mullen, B. C. Grant, J. DeLeon, J. C. Thrift, L. W. Chang, J. M. Irving, and D. J. Burke. 1994. Hybridoma stability. *Develop. Biol. Stand.* 83:55–64.

Chang, L. T., and R. P. Elander. 1986. Long-term preservation of industrially important microorganisms. in *Manual of industrial microbiology and biotechnology,* edited by A. L. Demain and N. A. Solomon. Washington, D.C.: American Society for Microbiology, pp. 49–55.

Coriell, L. L. 1979. Preservation, storage, and shipment. In *Methods in enzymology: Cell culture,* vol. 58, edited by W. B. Jakoby and I. H. Pastan. New York: Academic Press, pp. 29–36.

Crystal, R. G. 1995. Transfer of genes to humans: Early lessons and obstacles to success. *Science* 270:404–410.

Daggett, P. M., and F. P. Simione. 1987. *Cryopreservation manual.* Rochester, N.Y., USA: Nalge Company.

Darling, A. J., and J. J. Shapiro. 1996. Process validation for virus removal: Considerations for design of process studies and viral assays. *BioPharm* 9:42–50.

Del Tito Jr., B. J., M. A. Tremblay, and P. J. Shadle. 1996. Qualification of raw materials for clinical biopharmaceutical manufacturing. *BioPharm* 9:45–49.

Dunn, L., M. Abouelezz, L. Cummings, M. Navvab, C. Ordunez, C. J. Siebert, K. W. Talmadge. 1991. Characterization of synthetic macroporous ion exchange resins in low-pressure cartridges and columns. *J. Chromatog.* 548:165–178.

Eaton, L. C. 1989. Quantitation of residual *Escherichia coli* DNA in recombinant biopharmaceutical proteins by hybridization analysis. *J. Pharm. Biomed. Anal.* 7:633–638.

EC AD Hoc Working Party on Biotechnology/Pharmacy. 1991. *Validation of virus removal and inactivation procedures.* Note for Guidance, Draft #7. Luxembourg: Commission of the European Communities.

FDA. 1985. *Points to consider in the production and testing of new drugs and biologicals by recombinant DNA technology.* Bethesda, Md., USA: Food and Drug Administration, Center for Drugs and Biologics, Office of Biologics Research and Review.

FDA. 1987a. *Guideline on general principles of process validation.* Bethesda, Md., USA: Food and Drug Administration, Center for Drugs and Biologics and Center for Devices and Radiological Health.

FDA. 1987b. *Guidelines on validation of the limulus amebocyte lysate test as an end-product endotoxin test for human and animal parenteral drugs, biological products, and medical devices.* Bethesda, Md., USA: Food and Drug Administration, Center for Biologics Evaluation and Research and Center for Drug Evaluation and Research.

FDA. 1987c. *Points to consider in the characterization of cell lines used to produce biologicals.* Bethesda, Md., USA: Food and Drug Administration, Center for Drugs and Biologics, Office of Biologics Research and Review.

FDA. 1992. *Supplement to the points to consider in the production and testing of new drugs and biologicals produced by recombinant DNA technology: Nucleic acid characterization and genetic stability.* Bethesda, Md., USA: Food and Drug Administration, Center for Drugs and Biologics.

FDA. 1993. *Points to consider in the characterization of cell lines used to produce biological products.* Bethesda, Md., USA: Food and Drug Administration, Center for Drugs and Biologics, Office of Biologics Research and Review.

FDA. 1996a. *Addendum to the points to consider in human somatic cell and gene therapy (1991).* Bethesda, Md., USA: Food and Drug Administration, Center for Biologics Evaluation and Research.

FDA. 1996b. *FDA guidance concerning demonstration of comparability of human biological products, including therapeutic biotechnology-derived products.* Bethesda, Md., USA: Food and Drug Administration, Center for Biologics Evaluation and Research.

FDA. 1996c. *Guidance for industry for the submission of chemistry, manufacturing and controls information for a therapeutic recombinant DNA–derived product or a monoclonal antibody product for in vivo use.* Bethesda, Md., USA: Food and Drug Administration, Center for Biologics Evaluation and Research and Center for Drug Evaluation and Research.

FDA. 1996d. *Revised precautionary measures to reduce the possible risk of transmission of Creutzfeldt-Jakob disease (CJD) by blood and blood products.* Bethesda, Md., USA: Food and Drug Administration, Center for Biologics Evaluation and Research.

FDA. 1997a. *Points to consider in the manufacture and testing of monoclonal antibody products for human use.* Bethesda, Md., USA: Food and Drug Administration, Center for Biologics Evaluation and Research.

FDA. 1997b. *Proposed approach to regulation of cellular and tissue-based products.* Bethesda, Md., USA: Food and Drug Administration, Center for Biologics Evaluation and Research.

FR. 1993. Letter to manufacturers of drugs, biologics, and medical devices. *Federal Register* 59:44592.

FR. 1996. International Conference on Harmonization; Final guideline on quality of biotechnological products: Analysis of the expression construct in cells used for production of r-DNA–derived protein products. *Federal Register* 61:7006.

Gabler, R. 1986. Principles of tangential flow filtration. In *Filtration in the pharmaceutical industry,* edited by T. Meltzer. New York: Marcel Dekker, pp. 1–30.

Gherna, R. L. 1981. Preservation. In *Manual of methods for general bacteriology,* edited by P. Gerhardt. Washington, D.C: American Society for Microbiology, pp. 208–217.

Glazer, A. N., and H. Nikaido. 1995. *Microbial biotechnology: Fundamentals of applied microbiology.* New York: W.H. Freeman and Company.

Graham, F., L. J. Smiley., W. C. Russell, and R. Nairn. 1977. Characteristics of a human cell line transformed by DNA from human adenovirus type 5. *J. Gen. Virol.* 36:59–72.

Grun, J. B., E. M. White, and A. F. Sito. 1992. Viral removal/inactivation by purification of biopharmaceuticals. *BioPharm* 5:22–30.

Haaland, P. D. 1989. *Experimental design in biotechnology.* New York: Marcel Dekker.

Hageman, T. C. 1991. An analysis of clearance factor measurements performed by spiking experiments. *BioPharm* 4:39–41.

Hay, R. J. 1989. Preservation and characterization. In *Animal cell culture: A practical approach,* edited by R. I. Freshney. Oxford, UK: IRL Press, pp. 71–112.

Holmer, A. F. 1997. *New drug approvals in 1996.* Document #3002. New York: Pharmaceutical Research and Manufacturers of America.

ICH. 1995. *Viral safety evaluation of biotechnology products derived from cell lines of human or animal origin.* Step 2 Document. Geneva: International Conference on Harmonisation.

Johansson, B.-L. 1992. Determination of leakage products from chromatographic products aimed for protein purification. *BioPharm* 5:34–37.

Jolly, D. 1994. Viral vector systems for gene therapy. *Cancer Gene Therapy* 1:51–64.

Jones, A. J., and J. V. O'Connor. 1985. Control of recombinant DNA produced pharmaceuticals by a combination of process validation and final product specifications. *Develop. Biol. Stand.* 66:175–180.

Kelley, B. D., P. Jennings, R. Wright, and C. Briasco. 1997. Demonstrating process robustness for chromatographic purification of a recombinant protein. *BioPharm* 10:36–47.

Kirsop, B. 1987. Maintenance of yeast cultures. In *Yeast biotechnology,* edited by D. R. Berry, I. Russell, and G. G. Stewart. London: Allen & Unwin, pp. 3–52.

Kirsop, B. 1991. Service collections: Their functions. In *Maintenance of microorganisms and cultured cells,* edited by B. E. Kirsop and A. Doyle. London: Academic Press, pp. 5–20.

Kruse, R. H., W. H. Puckett, and J. H. Richardson. 1991. Biological safety cabinetry. *Clin. Microbiol. Rev.* 4:207–241.

Kung, V. T., P. R. Panfili, E. L. Sheldon, R. S. King, P. A. Nagainis, B. Gomez Jr., D. A. Ross, J. Briggs, and R. F. Zuk. 1990. Picogram quantitation of total DNA using DNA-binding proteins in a silicon sensor-based system. *Anal. Biochem.* 187:220–227.

Leahy, T. J. 1986. Microbiology of sterilization processes. In *Validation of aseptic pharmaceutical processes,* edited by F. J. Carleton and J. P. Agalloco. New York: Marcel Dekker, Inc., pp. 253–277.

Lonsdale, H. K. 1982. The growth of membrane technology. *J. Membrane Sci.* 10:81–181.

Lubinecki, A. S., M. E. Wiebe, and S. E. Builder. 1990. Process validation for cell culture-derived pharmaceutical proteins. In *Large scale mammalian cell culture technology,* edited by A. S. Lubinecki. New York: Marcel Dekker, pp. 515–541.

Lubinecki, A. S., K. Anumula, J. Callaway, J. L'Italien, M. Oka, B. Okita, G. Wasserman, D. Zabriskie, R. Arathoon, S. Builder, R. Garnick, M. Wiebe, and J. Browne. 1992. Effects of fermentation on product consistency. *Develop. Biol. Stand.* 76:105–115.

Mason, R. L., R. F. Gunst, and J. L. Hess. 1989. *Statistical design and analysis of experiments.* New York: J. Wiley.

McGregor, W. C., ed. 1986. *Membrane separations in biotechnology.* New York: Marcel Dekker.

Minobe, S., S. Tadashi, S. Tetsuya, and J. Chibata. 1983. Characteristics of immobilized histamine for pyrogen adsorption. *J. Chromatog.* 262:193–198.

Mossinghoff, G. J. 1996. *1996 survey: 284 biotechnology products in testing.* Document #3304. New York: Pharmaceutical Research and Manufacturers of America.

Muzyczka, N. 1992. Use of adeno-associated virus as a general transduction vector for mammalian cells. *Curr. Top. Microbiol. Immunol.* 158:97–129.

Naglak, T. J., M. G. Keith, and D. R. Omstead. 1994. Validation of fermentation processes. *BioPharm* 20:28–36.

Nolan, J. G., J. J. McDevitt, and G. S. Goldman. 1975. Endotoxin binding by charged and uncharged resin. *Proc. Soc. Exp. Biol. Med.* 149:766–770.

Novitsky, T. J., and M. J. Gould. 1985. Inactivation of endotoxin by polymixin B. In *Depyrogenation.* Technical Monograph No. 7. Bethesda, Md., USA: Parenteral Drug Association.

Onions, A. H. S. 1983. Preservation of fungi. in *The filamentous fungi,* vol. 4, edited by J. E. Smith, D. R. Berry, and B. Kristiansen. London: Edward Arnold, pp. 373–390.

Orkin, S. H., and A. G. Motulsky. 1995. *Report and recommendations of the panel to assess the NIH investment in research on gene therapy.* National Institutes of Health: http://www.nih.gov/news/panelrep.html.

Ostrove, J. M., W. Walsh, D. Vacante, and N. Patel. 1992. Molecular hybridization techniques and polymerase chain reaction (PCR) as methods for safety assessment of animal cells used in biopharmaceutical production. In *Animal cell technology: Developments, processes and products,* edited by R. E. Spier, J. B. Griffiths, and C. MacDonald. Oxford, UK: Butterworth-Heinemann Ltd., pp. 689–695.

PDA Biotechnology Task Force on Purification and Scale-up. 1992a. Industry perspective on the validation of column-based separation processes for the purification of proteins. *J. Paren. Sci. Tech.* 46:87–97.

PDA Biotechnology Task Force on Purification and Scale-Up. 1992b. Industry perspective on the validation of tangential flow filtration in biopharmaceutical applications. *J. Paren. Sci. Tech.* 46:S1–S15.

PDA Supplier Certification Task Force. 1989. Supplier certification–A model program. *J. Paren. Sci. Tech.* 43:151–157.

Per, S. R., C. R. Aversa, and A. F. Sito. 1989. Quantitation of residual DNA in biological products. *Clin. Chem.* 35:1859–1860.

Pirt, S. J. 1975. *Principles of microbe and cell cultivation.* Oxford, UK: Blackwell Scientific Publications.

PMA Computer System Validation Committee. 1986. Validation concepts for computer systems used in the manufacture of drug products. *Pharm. Technol.* 10:24–34.

Rigby, P. W. J., M. Dieckmann, C. Rhodes, and P. Berg. 1977. Labeling deoxyribonucleic acid to high specific activity in vitro by nick translation with DNA-Polymerase I. *J. Mol. Biol.* 113:237–251.

Riggin, A., and G. C. Davis. 1996. Reassessing the control of residual DNA in biopharmaceuticals. *BioPharm* 9:36–41.

Ross, G., R. Erickson, D. Knorr, A. G. Motulsky, R. Parkman, J. Samulsky, S. E. Straus, and B. R. Smith. 1996. Gene therapy in the United States: A five year status report. *Hum. Gene Therapy* 7:1781–1790.

Schindler, R., and C. A. Dinarello. 1990. Ultrafiltration to remove endotoxins and other cytokine-inducing materials from tissue culture media and parenteral fluids. *Bio Techniques* 4:408–413.

Scott, T., and E. H. Melvin. 1953. Determination of dextran with anthrone. *Anal. Chem.* 25:1656–1661.

Seamon, K. B. 1992. Genetic and biochemical factors affecting product consistency: Introduction to the issues. *Develop. Biol. Stand.* 76:63–67.

Seely, R. J., H. D. Wight, H. H. Fry, S. R. Rudge, and G. F. Slaff. 1994. Biotechnology product validation, part 7: Validation of chromatography resin useful life. *BioPharm* 7:41–48.

Smith, D. 1988. Culture and preservation. In *Filamentous fungi,* edited by D. L. Hawksworth and B. E. Kirsop. Cambridge, UK: Cambridge University Press, pp. 75–161.

Tsai, A., E. Englert, and E. Graham. 1990. Study of the dynamic binding capacity of two anion exchangers using bovine serum albumin as a model protein. *J. Chromatog.* 504:89–95.

Verma, I. M., and N. Somia. 1997. Gene therapy: Promises, problems and prospects. *Nature* 389:239–242.

Werner, R. G., H. Langlouis-Gau, F. Walz, H. Allgaier, and F. Hoffman. 1988. Validation of biotechnological production processes. *Drug Res.* 38:855–862.

WHO Study Group. 1987. Acceptability of cell substrates for production of biologicals. *WHO Technical Report Series* 747:1–29.

WHO. 1997. *Report of a WHO consultation on medicinal and other products in relation to human and animal transmissible spongiform encephalopathies.* Geneva: World Health Organization.

Wiebe, M. E., and S. E. Builder. 1994. Consistency and stability of recombinant fermentations. *Develop. Biol. Stand.* 83:45–54.

Wiebe, M. E., and L. H. May. 1990. Cell banking. In *Large-scale mammalian cell culture technology,* edited by A. S. Lubinecki. New York: Marcel Dekker, pp. 147–160.

6

ENSURING VIROLOGICAL SAFETY OF BIOLOGICALS: VIRUS REMOVAL FROM BIOLOGICAL FLUIDS BY FILTRATION

Hazel Aranha-Creado

Pall Corporation

THE NEED TO ENSURE THE SAFETY OF BIOLOGICALS

Biologicals are being increasingly used to treat and correct an array of medical conditions and metabolic dysfunctions. Table 6.1 is a listing of the general classes of biologicals in prophylactic and therapeutic use. Biologicals may be derived from mammalian fluids; for example, plasma-derived coagulation factors or immunoglobulins, hormones sourced from human urine, animal-derived products such as collagen and heparin, and antibodies produced in ascites. Tissue-derived products include hormones, e.g., human growth hormone (hGH), sourced from the pituitary gland of human cadavers, placenta-derived products such as albumin and bovine collagen, and other derivatives. With the prodigious advances in recombinant deoxyribonucleic acid (DNA) technology, a variety of biologicals, such as monoclonal antibodies (MAbs), cytokines, hormones, and

Table 6.1. Kinds of Biologicals in Prophylactic and Therapeutic Use

Type	Examples
Products derived from human/animal tissues/ fluids	• Blood/plasma derivatives, e.g., immunoglobulins, coagulation factors, antibodies produced in ascites, hormones from human urine • Tissue derivatives, e.g., hGH sourced from pituitary glands from cadavers, placenta-derived products such as albumin, bovine collagen
Products derived in human and animal cell lines	• Monoclonal antibodies, cytokines, hormones, recombinant proteins
Products prepared with/derived from microorganisms	• Bacterial vaccines and toxoids • Recombinant proteins expressed in bacterial or yeast systems
Products prepared with/ derived from viruses	• Viral vaccines utilizing inactivated and live attenuated viruses • Recombinant proteins expressed in viral systems, e.g., baculovirus expression systems

other recombinant proteins, are currently being produced in mammalian cell lines. Historically, bacterial and viral vaccines have been used as immunogens; similarly, bacterial toxoids have also been used to induce an antibody response in the recipient. More recently, microbial (bacterial/yeast) and viral systems are being used to express several recombinant proteins.

When biologicals are derived from systems that include a mammalian component (e.g., direct sourcing from these systems or supplementation with components from mammalian systems), the inherent risks concomitant with their use must be recognized.

Consequently, clinical acceptability of biologicals must, of necessity, be guided by risk-benefit analysis.

Historical Perspective

Historically, the utilization of biologicals commenced over a thousand years ago, when tribal customs included ingestion of animal- and human-derived tissues and fluids for religious and medical purposes. Cannibalistic practices, reportedly, contributed to the spread in New Guinea of Kuru disease, a neurodegenerative disease, similar to the present-day transmissible spongioform encephalopathies. The earliest medical application of the administration of biologicals was introduced by Edward Jenner in the late 18th century when calf lymph and human lymph from infected donors were administered to induce immunity to smallpox. Not surprisingly, due to the uncontrolled nature of the source material, infectious disease was a consequence of this practice. Similarly, in the late 19th century, while significant benefits were derived from rabies vaccine administration, clinical accidents were traced to the incomplete inactivation of the live attenuated virus used for vaccination. Literature on the benefits and associated hazards of immunization has been reviewed by Wilson (1967).

Examples of the transmission of viral agents by contaminated biologicals are presented in Table 6.2. Viral iatrogenic accidents have occurred due to one or more of the following reasons:

- Endogenous contaminants associated with the system used for virus cultivation, for example, the presence of SV-40 in the primary Rhesus monkey kidney cell line used for polio vaccine production (Shah and Nathanson 1976), and avian leukosis virus contamination of the hen eggs used for cultivation of yellow fever virus (Harris et al. 1966; Wilson 1967). Fortunately, in both these cases, long-term studies to evaluate the incidence of cancer in the recipients of these vaccines, i.e., polio vaccine (Fraumeni et al. 1970) and the yellow fever vaccine (Waters et al. 1972), have failed to establish a causal relationship.

- Contaminants have also been introduced when stabilizers or excipients such as serum are added. Human serum added to the yellow fever virus vaccine during the early years of World War II was subsequently demonstrated to have been contaminated with the hepatitis B virus (HBV) (Fox et al. 1942).

Table 6.2. Examples of Viral Iatrogenic Accidents Associated with Administration of Biologicals

Biological	Viral Contaminant	Source of Contaminant	Reference
Yellow fever vaccine	Hepatitis B	Albumin used as stabilizer	Fox et al. (1942); Wilson (1967)
FMD vaccine	FMD virus	Incomplete virus inactivation	King et al. (1981)
Polio vaccine	SV-40	Infected cell line	Shah and Nathanson (1976)
Polio vaccine	Polio virus (Cutter incident)	Incomplete virus inactivation	Nathanson and Langmuir (1963)
Rabies vaccine	Rabies virus (Fortaleza incident)	Incomplete virus inactivation	Para (1965)
Clotting factors (plasma-derived)	HIV; hepatitis A, B, C; parvovirus B-19	Infected donor; inadequate viral clearance during manufacturing	Aach and Kahn (1980); Centers for Disease Control (1996); Darby et al. (1995); Wang et al. (1994); Williams et al. (1990)
Growth hormone	Creutzfeld–Jakob disease[1]	Pituitary glands sourced from cadavers	Brown et al. (1992)
Immunoglobulins (plasma-derived)	Hepatitis C	Infected donor; inadequate viral clearance during manufacturing	Lever et al. (1984); Ochs et al. (1985)

[1]The identity of the etiological agent, i.e. virus (Manuelidis et al. 1987; Manuelidis 1994/prion (Prusiner 1984), has not been established to date.

- Incomplete inactivation of the viral immunogen was the causative factor in incidents such as the Cutter incident with polio vaccine (Nathanson and Langmuir 1963), the Fortaleza accident in Brazil with the rabies vaccine (Para 1965), and the foot-and-mouth disease (FMD) vaccine (King et al. 1981). Even though epidemiological studies implicated the FMD and yellow fever vaccines in the transmission of the viral agent, actual demonstration/correlation with the vaccine was facilitated only recently, as a result of advances in molecular diagnostic techniques.

The considerable improvement in the safety record of vaccines is a reflection of the significant strides made in virus cultivation and detection methodologies combined with rigorous and effective process controls. Though the risk of a therapeutic biological being the vector for viral disease transmission has decreased considerably, viral dissemination via administration of biologicals continues to occur. For example, there have been several recent reports of the transmission of hepatitis A (Centers for Disease Control 1996; Lawlor et al. 1996), hepatitis B (Aach and Kahn 1980), hepatitis C (Wang et al. 1994), human immunodeficiency virus (HIV) (Darby et al. 1995), and parvovirus (Brown and Young 1997; Williams et al. 1990) infections. In all of these cases, the infection was traced to the use of contaminated source materials (blood supplies).

Contemporary Approach

A discussion of the virological safety of biologicals requires the consideration of several complex issues, many of which are yet to be refined and resolved. Safety specifications are clearly defined in the case of bacterial removal (i.e., demonstration of the absence of bacterial contaminants), and these specifications have been implemented and validated over the last several decades. However, establishing absolute virological safety poses a problem for several reasons. Some of the concerns stem from the following considerations: the great diversity of viruses and, consequently, the necessity of performing specific assays for each virus; the lack of sensitivity of methods, i.e., difficulty in detecting low levels of infectious units that may be of medical concern; the extremely efficient amplification of viruses; and the unknown amphitropism of unknown viral variants. Direct testing for the absence of viral contamination from a finished product is not considered sensitive enough for establishing freedom from infectious virus.

In addition to conventional viruses, it is becoming increasingly clear that the potential for transmitting unconventional agents via the administration of biologicals must also be considered. The transmissible spongiform encephalopathies (TSEs) are a group of fatal neurodegenerative diseases with a long preclinical phase; these diseases are marked by a characteristic vacuolation or sponginess in the affected brain tissue. Examples of TSEs in humans include kuru (an infectious disease propagated due to anthropophagous practices), Creutzfeld–Jakob disease (CJD), which may take either an infectious, genetic, or sporadic form; and Gerstmann-Straussler-Scheinker (GSS) disease and fatal familial insomnia (FFI), which are rare familial disorders. Cases of iatrogenic CJD have resulted from neurosurgical procedures associated with tissue implantation, e.g., dura mater grafts; and treatments with hormones such as hGH derived from cadaver tissue. Thus far, CJD transmission via blood or blood derivatives has not been demonstrated. Other TSEs of concern include scrapie in sheep and bovine spongiform encephalopathy (BSE) or "mad cow disease" in cattle. While the TSE infective agents demonstrate species specificity (i.e., they are unable to initiate infection in a heterologous host), recent reports of atypical cases of CJD (new variant CJD, nvCJD) in the United Kingdom raised fears that BSE might be transmissible to man (Collinge et al. 1996).

The TSE infective agents are poorly defined, which makes their detection complex. All TSE isolates contain aggregates of a glycoprotein called proteinase K-resistant prion protein, also referred to as PrP-res or PrP^{Sc} for scrapie. There is also a "normal" prion protein, PrP^{C} (the normal cell protein), also referred to as PrP-sen as it is sensitive to proteinase K, which is present in the brain of all mammalian species. While PrP^{Sc} may be used as a biochemical marker for both the disease and the infective agent, the only available infectivity assay is a rodent bioassay that may take up to 18 months before results are available. To date, it has not been established whether the etiological agent involved in TSEs is a conventional slow virus (Manuelidis et al. 1987; Manuelidis 1994) or an unconventional agent, a nucleic acid-free, self-replicating infectious protein that is designated a prion (Prusiner 1984).

An absolute guarantee of viral safety for human/animal-sourced/derived products does not seem feasible at the present time. However, with a concerted approach involving the incorporation of multiple barriers to viral transmission, the margin of safety can be significantly increased. To date, there have been no definitive reports of adverse incidents associated with the administration of contaminated continuous cell line (CCL)–derived products. This is, however, no cause for complacency; it merely highlights an excellent manufacturing record and the necessity for continued incorporation of appropriate viral contamination control strategies.

This chapter will focus on the virological safety of CCL–derived products, with a specific emphasis on virus removal by filtration. While the safety and procedural considerations in the case of plasma-derived versus CCL–derived products are similar, special emphasis will be given to products derived from CCLs. Additionally, while a specific discussion of unconventional agents, such as those involved in TSEs, is not included, several of the issues and concerns are similar to those involved in virological safety, and similar strategies are recommended.

SOURCES OF VIRAL CONTAMINATION

Manufacturing processes utilize a myriad of raw materials in both production and purification processes; these have a significant potential for contributing to the viral load of a system. Briefly, viral contaminants of systems can originate from within the system (i.e., endogenous contaminants) or they may be introduced into the system as a consequence of additions and other manipulations (i.e., adventitious contaminants).

Endogenous Viral Contamination in Continuous Cell Lines

CCLs serve as substrates for the production of a number of biopharmaceuticals and, consequently, from a process standpoint, they constitute a manufacturing component. CCL characterization is an integral part of establishing quality control (QC) over the manufacturing process. In addition to ascertaining the identity and purity of the cell line, for example, by identifying unique reference markers, testing for the presence of endogenous or adventitious agents is also undertaken. CCLs used as substrates in the manufacturing of biopharmaceuticals are based on extensively characterized and controlled seed lot systems designated as the master cell bank (MCB) and the working cell bank (WCB). The absence of cytopathogenic effects in the seed lots cannot be construed as the absence of viral presence. Persistent latent infections in cell lines can remain undetected until appropriate immunological, cytological, ultrastructural, and/or biochemical tests are used. All cell banks must be tested at passage levels well beyond that to be used for production; they are tested for the presence of endogenous viruses by the induction of cells, with an agent such as iododeoxyuridine (IUdR), and examined by electron microscopy and for reverse transcriptase (RT) activity.

Molecular hybridization can also be used to exclude the presence of specific viral sequences (Magrath 1991).

In general, viral contaminants may originate from the animal from which the cell line is derived, in which case they may be cytopathic, such as SV-5 contaminants of primary monkey kidney cells; chronic, such as the C-type particles of hybridoma lines used for the production of murine MAbs; or latent, such as integrated murine retroviruses. Cytolytic viruses, which were extremely common in the early days of vaccine production, are now of less concern because they are essentially self-excluding, because these cell lines would not survive to the production stage if infected with such agents. Viral contamination is more problematic where no overt cytopathic effects are observed, as in the case of latent or chronic viruses, either intrinsic to the cell type or introduced by the culture methods used in their production. Table 6.3 lists some of the viruses that could potentially be associated with nonhuman and human-derived cell lines.

One concern in several systems is the existence of retroviral proviruses. Retroviruses replicate their ribonucleic acid (RNA) genome by viral-encoded reverse transcriptase; this replication is followed by integration of the reverse-transcribed DNA into the host chromosomal DNA to produce a provirus to be used for transcription. As with chromosomal DNA, retroviruses are transmitted vertically (from parents to offspring) through germ cells. All vertebrates are considered to have endogenous retroviral genomes (Ono 1988). While the direct biological consequences of the presence of endogenous retroviruses have not been demonstrated, they are a theoretical safety concern of biologics produced in continuous cell substrates. The putative risk stems from their morphological and biochemical resemblance to tumorigenic retroviruses.

The presence of endogenous viruses in rodent-derived cell lines is well documented. In addition to type A retroviral particles observed intracellularly in a variety of cells, infectious and noninfectious type C retroviral particles have been reported; these type C particles have been visualized by electron microscopy but may be undetectable by viral infectivity and/or RT assays. For example, Chinese hamster ovary (CHO) cells used in the expression of recombinant proteins have endogenous viruses (Anderson et al. 1991; Emanoil-Ravier et al. 1991; Lueders 1991). Similarly, syrian hamster cells have been demonstrated to contain intracisternal R-type and A-type particles, with no accompanying reverse transcriptase or infectivity activity (Bergmann and Wolff 1981). Some of the endogenous and latent viruses of relevance are included in Table 6.3.

Table 6.3. Viruses of Concern in Cell Production Work

Viruses potentially associated with nonhuman cell lines

- Hantavirus
- Lymphocytic choriomeningitis (LCM) virus
- Ectromelia virus
- Murine hepatitis
- Simian viruses
- Sendai viruses
- Avian leukosis viruses
- Bovine viral diarrhea virus (BVDV)

Viruses potentially associated with human cell lines

- Retroviruses
- Hepatitis viruses
- Human herpes viruses
- Cytomegalovirus
- Human papilloma virus
- Epstein-Barr virus

Monoclonal antibodies produced in human cell lines are less likely to induce adverse immunological responses in the recipient; however, they raise unique considerations from a virological safety standpoint. The cells of interest obtained from a human donor are the B lymphocytes, which can harbor several viruses, including retroviruses, hepatitis viruses, human herpes viruses, cytomegalovirus, and human papilloma virus (Table 6.3).

Potential for Association of Viral Zoonoses with Continuous Cell Lines

Hazards may also be posed by zoonotic agents, which may be difficult to detect as contaminants in cell lines and, consequently, have the potential of contaminating biological products. The importance of rodent zoonotic agents has increased in the last few decades in view of the widespread use of murine cell lines in the manufacture of MAbs.

Hantaan virus, lymphocytic choriomeningitis (LCM) virus, and the rat rotavirus are well-documented zoonotic viruses (Mahy et al. 1990). Hantaan virus, the causative agent of Korean hemorrhagic fever, replicates in tumor and other cell lines; outbreaks of this disease in individuals exposed to infected colonies of laboratory rats have been reported in several countries (LeDuc 1987). Similarly, infection of transplantable tumor lines with LCM virus from host animals reportedly led to an outbreak of disease (Bowen et al. 1975). Although some rodent viruses have not yet been recorded as being transmissible to humans under the normal conditions of contact between humans and rodents (i.e., exposure under laboratory conditions), this does not rule out the possibility of their pathogenicity to humans. Introduction through a nonconventional route, for example, inadvertent administration of the virus intravenously, or when attached to lymphocytes reintroduced into patients after in vitro procedures, could give rodent viruses access to cell types, which they would not normally encounter at mucosal surfaces (the normal route of transmission of zoonoses), and thus allow their replication (Carthew 1989).

Safety validation programs performed on the seed lot of the CCL must, therefore, also include testing for extraneous viruses, including LCM virus (when rodent originated), and other source species viruses, in the case of cells of nonrodent origin (Dorpema 1988).

Adventitious Virus Contamination of Continuous Cell Lines and Manufacturing Processes

Contaminants may also be derived from the raw materials/additives used in the production of serum, trypsin, or other materials of human/animal origin. Table 6.4 lists some of the adventitious viruses that could possibly contaminate biologicals. For example, certain viral agents, such as Epstein-Barr virus and Sendai virus, have been used for the establishment or transformation of cells. Viruses may be introduced along with culture media or other supplements. Bovine viral diarrhea virus (BVDV), infectious bovine rhinotracheitis (IBR) virus, and parainfluenza-3 virus (PI-3) have been reported in unprocessed and commercial lots of fetal bovine serum; of these, BVDV is of primary concern because IBR and PI-3 are relatively large and would most likely be removed by standard sterile filtration processes

Table 6.4. Potential Sources of Adventitious Viral Contamination

Source of Viral Contamination	Examples of Viruses
Virus used for induction of expression of specific genes encoding a desired protein	Epstein-Barr virus, Sendai virus, other inducing agents
Reagents/additives used during production, e.g., serum, culture media, trypsin, growth factors, and other supplements	BVDV, IBR virus, parainfluenza-3
Reagents used during purification, e.g., affinity columns (MAbs) for purification	Viruses from MAbs/ unknown viruses from large animal polyclonal antibodies
Excipients used during formulation, e.g., serum	Human viruses such as hepatitis B
Manufacturing facility/personnel	Rhinovirus, respiratory syncytial virus, rotaviruses

employed during manufacture (Erickson et al. 1991). Contamination incidents where epizootic hemorrhagic disease virus (EHDV) was associated with CHO cells (Rabenau et al. 1993), and contamination of unprocessed bulk harvest with a parvovirus (minute virus of mice) has been reported (Garnick 1996). Viruses may also be introduced during purification processes. For example, affinity chromatography using MAbs as ligands for capture of coagulation factors or other biological molecules is commonplace (Lawrence 1993). Use of these biologically derived reagents increases the potential for the introduction of adventitious viral agents into the product. Excipients or stabilizers used during product formulation can contribute to the viral load; for example, the serum used as the excipient in the production of the yellow fever vaccine (administered during World War II) was implicated as the source of the contaminating hepatitis B virus in the vaccine (Wilson 1967). The possibility of the introduction of viruses such as rhinovirus, respiratory syncytial virus, and rotaviruses from manufacturing environments or personnel must also be considered (Hay 1991).

Adventitious agents have been detected in nonhuman cell lines. It has been suggested that CHO cells, a cell line frequently used in

MAb production, may be contaminated with hantavirus. Ectromelia virus, a natural pathogen of mice, was shown to replicate in all murine lymphoma cells and a small proportion of the hybridomas tested (Buller et al. 1987).

Human cell line–derived MAbs pose enhanced safety concerns because the absence of a species barrier compounds the possibility of iatrogenic viral transmission. Human cell lines are often derived from human/mouse and human/human hybrids, in which human cells are transformed by the Epstein-Barr virus (Glaser 1988). Epstein-Barr virus is a human herpes virus associated with Burkitt's lymphoma and infectious mononucleosis. In addition to the viral load of the cell line itself, and the deliberate introduction of viruses for the immortalization of the cell line, the methods involved in preparing the cell line for production also raise virological issues. For example, the B lymphocytes are isolated by rosetting with sheep erythrocytes, introducing possible contamination by ovine agents such as Visna-Maedi virus. Additionally, because of the lack of a suitable human myeloma cell line, the human antibody-producing cell line may be established by fusing with a murine myeloma line or a preexisting hybrid between a human lymphocyte and a murine myeloma line. This fusion may lead to activation of infectious murine leukemia viruses (Minor 1991). Appropriate sourcing in terms of the origin of the human cell line and details regarding its development and appropriate validation of production processes are key considerations in ensuring their safety.

DETECTION OF VIRAL CONTAMINATION

Several qualitative and quantitative methods exist for the detection of viruses; Table 6.5 lists some of the methods. Typically, biological assays (i.e., infectivity assays) are the assays of choice. Data obtained from other detection methods may be used to supplement, not supersede/replace, infectivity assays. However, in cases where susceptible cell lines for virus cultivation are unavailable, for example, hepatitis C virus, data from assays such as the polymerase chain reaction (PCR) are deemed acceptable.

Biological (Infectivity) Assays

Virus infectivity may be evaluated either by inoculation into susceptible laboratory animals (i.e., in vivo assays) or by in vitro inoculation of susceptible cell lines. Cell cultures may be monitored by observation of cytopathic effects (CPEs), such as the formation of

Table 6.5. Examples of Tests Used for Detection of Viruses

Type of Assay	Examples	Capability	Limitations
Biological Assays	In vitro infection of cell cultures, followed by observation of cytopathic effects	Quantitative	Culture-specific infectivity/replication; therefore, different assay system for each virus
	In vivo infection of susceptible animals	Quantitative	
Antibody production tests	Induction of antibodies in susceptible animals, (e.g., mouse, rat, hamster) followed by detection of antibodies to specific viruses	Qualitative	Host-specific infectivity and immune response
Biochemical assays	Reverse transcriptase assays, radiolabel incorporation into nucleic acids, immunofluorescence, Western blots, radioimmunoassays	Semiquantitative	Detects enzymes with optimal activity under preferred conditions; also, interpretation may be difficult due to presence of cellular enzymes
Molecular probes	Polymerase chain reaction (PCR), nucleic acid hybridization assays	Semiquantitative	Primer sequences must be present; detects non-infectious virus
Morphological assays	Transmission electron microscopy	Qualitative	Assessment of identity only; detects noninfectious virus

plaques, focus forming units, or induction of abnormal cellular morphology, or by evaluation of hemagglutination capacity when grown in other cell lines. In vivo assays may be necessary to detect viruses that do not grow in cell cultures; detection methods include infection of embryonated eggs or inoculation into susceptible laboratory animals, such as suckling and adult mice. Infectivity assays are quantitative and must be used whenever possible. However, the requirement of a different assay system for each virus due to cell culture specific infectivity makes biological assays cumbersome.

Antibody Production Tests

Species specific viruses present in rodent cell lines may be evaluated by inoculation into virus-free animals and quantification of serum antibody levels or enzyme activity after a specified time period. This method has been used for the testing of murine hybridoma lines and other rodent lines such as CHO cells. Mouse antibody production (MAP) tests, hamster antibody production (HAP) tests, and rat antibody production (RAP) tests have been used to detect viruses such as Hantaan virus, LCM virus, Sendai virus, and reovirus type 3 (Reo-3), which have the potential for infecting humans and other primates. In addition to its qualitative nature, the labor intensivity and the host specific response associated with the antibody production test does not make it amenable for routine assay purposes.

Biochemical Assays

Biochemical assays such as RT assays, radiolabel incorporation into nucleic acids, radioimmunoassays, immunofluorescence, and Western blots have been used for virus detection. For example, RT assays are used to detect retroviruses (Anderson et al. 1991). However, these tests are semiquantitative; they also detect enzymes with optimal activity under the test conditions, but their interpretation may be difficult due to the presence of cellular enzymes or other background material.

Molecular Probes

Molecular probes such as hybridization assays or PCR assays are becoming increasingly popular because of their specificity and the rapidity of the results. However, these methods detect only the presence of nucleic acid (DNA/RNA) and cannot differentiate between infectious or noninfectious particles. Additionally, the method is applicable only when the genomic sequence of the virus is known, as in the case of retroviral genomes. PCR is especially relevant either if the viral agent cannot be grown in vitro, e.g., type A retroviral particles, or for viruses such as hepatitis B and C where there are severe limitations to culturing them in vitro. While PCR has an obvious role in the safety testing of biologicals, whenever possible, it should be used to complement data obtained from infectivity assays.

Morphological Assays

Morphological assays such as electron microscopy, while of limited value to assay culture fluids, may be used, for example, to examine the cells themselves for intracellular viral particles (Anderson et al. 1991). Electron microscopy is applicable for cell assessment; however, it is of questionable value for the quantitation of virus particles (Liptrot and Gull 1991). As with molecular probes, electron microscopy is incapable of determining the biological relevance of a particle in terms of its infectivity. This is a common situation in hybridoma cultures where type A retroviral particles are detected in large numbers (as determined by electron microscopy) but infectivity assays show little or no infectivity.

Summary

In order to assess the effectiveness of the viral clearance process, the ability to quantitate the amount of virus is essential. In general, the methods used must accurately and reproducibly quantitate the viral load. The effectiveness of viral clearance is determined by comparing the virus concentration prior to treatment to the concentration of virus in the sample after treatment. The methods of choice are quantitative infectivity assays, such as the detection of plaque formation or other CPEs, such as syncytia or foci formation, endpoint titrations ($TCID_{50}$ assays), detection of virus antigen synthesis, or other methods. Viral detection methods should have adequate sensitivity and reproducibility and should be performed with sufficient replicates and controls to ensure adequate statistical accuracy of the test results.

It must be noted that while positive results are meaningful, negative results are ambiguous. This is because it is not possible to determine whether the negative result reflects inadequate sensitivity of the test for the specific virus, selection of a test system (host) with too narrow a specificity, poor assay precision, limited sample size, or, basically, absence of virus.

METHODS OF VIRAL CLEARANCE

A review of the current approaches used for the containment of viruses indicates that viral clearance may be achieved either during manufacturing processes as a consequence of routine processing and purification methods (i.e., serendipitously or fortuitously) or, alternatively, it may be effected by deliberate viral clearance technologies incorporated into the process. For example, serendipitous viral clearance includes steps in the purification process that have virus reduction potential, such as pH effects either during processing (Hamatainen et al. 1992) or when low pH buffers are used to elute proteins from chromatography columns (Grun et al. 1992); virus inactivation by reagents used in the purification process (Horowitz and Horowitz 1984); and fractionation, which can selectively partition certain viruses (Marcus-Sekura 1992; Morgenthaler and Omar 1993).

Based on the mechanism of action, viral clearance methods are classified as either removal or inactivation methods. Table 6.6 presents some of the methods currently in use. Virus removal methods include removal by size exclusion, for example, filtration (Aranha-Creado et al. 1997; DiLeo et al. 1993a and b), adsorption to certain matrices, e.g., chromatography (Burnouf 1993; Gomperts 1986), and partitioning into a different fraction (Marcus-Sekura 1992; Morgenthaler and Omar 1993). Some of the methodologies to inactivate viruses include heat (Murphy et al. 1993; Pasi and Hill 1989), photochemical inactivation (Prudouz and Fratantoni 1994), chemical inactivation (Dichtelmuller et al. 1993), ultraviolet (UV) inactivation (Prince et al. 1983), and solvent-detergent inactivation (Horowitz et al. 1985; Horowitz 1989; Mannucci and Colombo 1988). The inactivation strategy used will be dictated by the lability of the virus, the stability of the biological preparation, and the effect on other components in the preparation (Barrowcliffe 1993). While inactivation methods can be very effective in decreasing the viral burden, there are several limitations. Heat treatment can denature certain proteins (Gleeson et al. 1990; Prince et al. 1987). Processes such as solvent-detergent inactivation target a specific characteristic (i.e., presence of a lipid envelope) and are, consequently, effective only against enveloped viruses (Horowitz 1989; Mannucci 1992).

Viruses may vary with regard to their susceptibility to inactivation treatments such as heat or radiation; within a population of viruses, heat- or radiation-resistant fractions may exist. Additionally, the possibility of the production of neo-antigens and, thus, the induction of antibodies/inhibitors in the recipient must be considered (Peerlinck et al. 1993). Also, stabilizers are sometimes used during heat or solvent-detergent inactivation to ensure that the biological activity of the active moiety is not compromised. These agents need to be removed from the final product, which may affect product

Table 6.6. Methods Commonly Used for the Reduction of Viral Burden in the Manufacture of Biologicals and Biopharmaceuticals

Clearance Method	Examples	References
Virus removal	Filtration	Aranha-Creado et al. (1997); DiLeo et al. (1993a and b)
	Partitioning/ fractionation	Marcus-Sekura (1992); Morgenthaler and Omar (1993)
	Chromatography (ion exchange, affinity, gel filtration)	Burnouf (1993); Gomperts (1986)
Virus inactivation	Heat (heating in solution, heating of dry products, heating of dry products in steam, heating of dry products in organic solvents)	Murphy et al. (1993); Pasi and Hill (1989)
	Chemicals, e.g., organic solvents, detergents, β-propiolactone	Dichtelmuller et al. (1993); Horowitz (1989); Mannucci and Colombo (1988);
	Radiation (gamma, ultraviolet, microwave)	Prince et al. (1983)
	pH alterations	Grun et al. (1992); Hamatainen et al. (1992)

recovery (Heimburger et al. 1981). Furthermore, in addition to protein denaturation and the consequent loss of biological activity, the possibility also exists that viral inactivation methods could alter/ increase the antigenicity of both the active ingredient and other proteins in the product. For example, enhanced thrombogenicity, as a consequence of viral inactivation methods, has been reported with factor IX preparations (Prowse and Williams 1980).

For the last several decades, filtration has been successfully used to remove bacteria and mycoplasma from pharmaceuticals and biologicals. Filtration does not compromise the biological integrity of the product; also, there is no requirement for inclusion of stabilizers/additives, which have a potential for toxicity or may affect product recovery. Membrane filtration by size exclusion is the preferred mechanism because it provides a predictable mechanism of viral particle removal. It constitutes a "robust" mechanism for virus removal because it is not affected by minor alterations in process parameters. Other mechanisms that may influence virus removal by filtration include viral adsorption to the filter surface by electrokinetic interactions and controlled filtration conditions, such as flow rate, pressure, and temperature. Retention capacity may be enhanced due to effects such as gel polarization. It must be emphasized that concomitant with the requirement for adequate virus removal by size exclusion is the inherent necessity for no significant loss in product concentration and/or activity following filtration processes.

REGULATORY CONSIDERATIONS IN VIRAL CLEARANCE

In general, the major factors influencing the viral safety of biologicals are the following:

- The species of origin of the starting material, i.e., the host range for viral infectivity is dictated by the presence of specific cellular receptors. Nonhuman viruses are less likely to initiate infection in humans due to the species specificity of these viruses; thus, for example, most animal viral diseases are not transmissible to humans. With human-sourced products, e.g., blood coagulation factors, due to the absence of a species barrier, the probability of human infection by contaminating viruses is increased.

- The degree of source variability of starting material (e.g., human plasma-derived products that are manufactured from pooled donations and that cannot be controlled versus cell culture products derived from a well-characterized MCB) and the possibility of testing the source material for the presence of viral contaminants (feasible for blood donation but not feasible for animal-derived products).

- The nature of the process used for obtaining the product (purification procedures) and its capacity to eliminate viruses.

- The existence of specific steps for viral clearance included in the process (Horaud 1993).

In view of the unique considerations associated with viral contaminants (i.e., actual versus theoretical presence) and the limitations in assay methodologies, it is unlikely that it will be possible to provide absolute assurance of the absence of viral contaminants. Consequently, all regulatory agencies emphasize a holistic approach directed at risk minimization. Risk assessment relies completely on the quality and appropriateness of the available data and on adequate evaluation.

The minimization of risk of exposure to real and theoretical viral contaminants is achieved by incorporating multiple orthogonal barriers to virus transmission in an integrated manner, so as to provide overlapping and complementary levels of protection, with different modes of action, and separately derived virus reduction potential (Builder et al. 1989). Safety is thus the result of multiple barriers operating in concert. While each individual approach may have limitations, their use in an integrated manner provides overlapping and complementary levels of protection from known and potential viruses. These approaches represent the only currently feasible approach in the face of theoretical risks that cannot be adequately characterized by classical technology.

Table 6.7 summarizes the three approaches that have been recommended for risk minimization. Essentially, there are three aims of safety strategies:

1. To prevent access of a virus by screening of raw materials/precursors used (i.e., evaluating the starting materials used in the production process, e.g., the cell bank, the culture medium, serum supplements, etc.). The nature of the source material will reflect the degree of risk associated with the final product. Source materials such as human blood donations are routinely screened. Similarly, quality assurance (QA) on CCL–derived products includes documentation of the origin of the cells used and examination of the cell banks established. Certain source materials are potentially more hazardous than others because of their origin; thus hGH derived from recombinant bacteria raises essentially no virological concerns, while material derived from human cadavers raises many. Similarly, many bovine-derived pharmaceuticals are being replaced by their recombinant

Table 6.7. Overall Strategy to Prevent Viral Contamination

Barriers to entry, i.e., appropriate sourcing

- characterization of MCB and WCB

- selection and screening of donors

Incorporation of "robust" virus clearance steps

- serendipitous/deliberate virus removal/inactivation steps—filtration, chemical inactivation, etc.

Testing during production to ensure absence of contaminating virus

- virus detection assays such as infectivity assays and PCR during manufacture

---

equivalents because of the risk of BSE. Especially in the case of TSE–infective agents (the etiological agent for mad cow disease) appropriate sourcing is a key consideration because inactivation of these agents is difficult and the detection assays are extremely complex.

2. To monitor production, i.e., testing for endogenous and adventitious virus.

3. To evaluate the manufacturing process, i.e., process validation for viral clearance. Review of the clearance methodologies is essential to determine the "robustness" of the individual viral clearance steps. Certain steps such as centrifugation and precipitation, where minor variations in process parameters may alter virus removal, must be viewed with caution.

Establishing the absence of infectious virus in the final product often is not derived solely from direct testing for its presence; a demonstration that the purification regimen is capable of removal and/or inactivation of viruses is also necessary (FDA 1993, 1997; ICH 1997). Similar recommendations have also been made in Europe by the Committee for Proprietary Medicinal Products (CPMP) in the "Note for Guidance on Plasma-Derived Medicinal Products" (1997).

While the necessity for risk assessment and incorporation of not merely adequate but excess virus clearance capacity is acknowledged, the amount of excess capacity required has not been clearly defined.

It has been suggested that "the overall viral reduction should be greater than the maximum possible virus titer which could potentially occur in the source material" (CPMP 1991). The recommendations do not provide specifics with regard to the extent of excess viral clearance that would constitute an acceptable safety margin. However, in general, processes must be validated to remove or inactivate 3–5 orders of magnitude more virus than is estimated to be present in the starting material (Marcus-Sekura 1992). For products known to possess viral contaminants (e.g., endogenous retroviruses), it is necessary to determine the theoretical viral burden per dose equivalent of the biological product. Thus, in the case of CCL–derived products, an estimate of viral burden based on biological/biochemical (e.g., infectivity, RT) and morphological (electron microscope) assays is essential. Other investigators (Shadle et al. 1995) have recommended that three independent components be summed mathematically to define the number of logs of viral clearance needed for a given process:

$$\text{Required clearance} =$$
$$\text{risk} + \text{biological safety factor } (F_B) + \text{statistical safety factor } (F_S)$$

The magnitude of F_B will be higher in the case of source materials possessing measurable viral infectivity. Clearance estimates and their variances can be calculated for each orthogonal unit clearance operation and estimates can then be combined to form an interval estimate of overall process clearance capacity.

The contemporary approach applied by regulatory groups indicates that application of the guideline is not to be interpreted as a prescriptive set of requirements to gain licensure but rather must be interpreted in accordance with current standards of scientific methodology and rigor. Flexibility is necessary, resulting in different combinations of measures for each set of circumstances. For example, certain source materials known to be contaminated with viruses or virus-like particles (e.g., hybridoma cells used for the production of MAbs are frequently contaminated with type C murine retroviruses) are, nonetheless, deemed acceptable since the contaminating entity is considered to pose only a low risk to human recipients. Conversely, in the case of plasma-derived products, presence of even low levels of virus have been demonstrated to initiate infection, and, consequently, adequate clearance methodologies must be incorporated to ensure virological safety.

VALIDATION CONSIDERATIONS IN THE DOCUMENTATION OF VIRAL CLEARANCE

The need to validate all processes related to the manufacturing of biopharmaceuticals has been well established in the Good Manufacturing Practice (GMP) regulations and various guidelines. As per the U.S. Food and Drug Administration (FDA) in *Guidelines for Sterile Drug Products Produced by Aseptic Processing,* microbial retention must be demonstrated under simulated pharmaceutical conditions in order to document the performance claims of the filter (FDA 1987). Similar considerations apply in validation of manufacturing processes for virus removal (CPMP 1991; ICH 1997).

The primary objectives of documenting viral clearance (i.e., viral validation studies) are twofold: (1) to demonstrate the effective removal/inactivation of viruses that are present/could potentially exist and (2) to provide indirect evidence that the process conditions have adequate built-in safety measures to ensure elimination of theoretical/novel viruses that could potentially exist. Viral clearance validation must, therefore, be undertaken within the parameters set forth for the production process (i.e., process modeling must be accurate) and with a membrane that is representative of the membrane used in filter cartridge device manufacture. Validation for viral clearance essentially involves challenging (spiking) the product with high titers of infectious virus under conditions that simulate process parameters and quantitating the virus in pre- and posttreatment samples.

Some of the factors that affect the results of viral clearance studies include the appropriateness of the scaled-down version, the choice of "model" viruses used, the identification of process variables that may alter the efficacy of the virus elimination steps, and the performance of experiments that should be based on good virological practices (correct titrations based on sound statistical grounds, inactivation kinetics, balance for partitioning steps, etc.). The justification for and the extent of the required validation studies will vary depending on the type of product (e.g., plasma-derived versus a CCL–derived product) and the manufacturing process. These considerations are described in the next section.

Aspects of Virus Validation Studies

Some of the factors to be considered when conducting validation studies are discussed below.

Type of Viral Contaminants and Baseline Viral Load

Risk assessment to determine the kind of viral contaminant and the baseline loads, if any, must be undertaken. For example, as stated previously, in spite of routine blood screening, biologicals such as those blood sourced from multiple donors may occasionally be contaminated with virus (e.g., HIV; hepatitis A, B, C; parvovirus B-19). These viruses may have escaped detection by routine detection methods because of their presence at very low levels or because of blood donations by seronegative donors during the infectious "window period" of the disease when the donors were undergoing seroconversion (Lackritz et al. 1995; Schreiber et al. 1996). More recently, viruses such as hepatitis G have been reportedly transmitted by blood transfusions (Alter et al. 1997). Additionally, new viral variants transmissible by blood have been reported. For example, the presence of human herpes virus 8 (the etiological agent implicated in Kaposi's sarcoma in HIV–infected individuals) in a healthy blood donor raises the issue of transmission of this agent via blood transfusions (Blackbourn et al. 1997; Simmonds 1998). Conversely, the demonstrated presence of endogenous retroviruses in CCLs may pose only a theoretical safety concern as the biological relevance of their similarity to tumorigenic retroviruses has not been demonstrated. Thus, the baseline viral load and the potential pathogenicity of the contaminant viruses must be ascertained in order to assess the viral safety requirement for the process.

Scale-Down Considerations

Scaling down of the production process and validating that the scaled-down process corresponds to the production process, i.e., process modeling, must be accurate. In view of logistic and GMP considerations, such as not introducing viral contaminants into the production environment, validation studies, are, of necessity, scaled down. The products generated by the large- and small-scale processes should be equivalent in terms of purity, potency, and yield. The scaled-down process should simulate production parameters in terms of temperature, buffers, linear flow rates, and so on. The effectiveness of each process step should be evaluated, and the results used in setting appropriate in-process limits.

Selection of Test Viruses

Selection of the test virus is influenced by the relevance and relationship of the test virus to the viruses of risk, the possibility of growing them to high titers in vitro, and the availability of accurate infectivity assays. Additionally, it is necessary to ensure that the product fluid used in the validation testing is free of antibodies to the test viruses. This is potentially a problem with viruses such as herpes simplex virus type 1 (HSV-1) and poliovirus. Therefore, especially when working with plasma-derived biologicals, the precaution of screening the starting material for neutralizing antibody before initiating the viral validation experiments is recommended. Failure to do so may result in the generation of artificially high clearance factors for a process, resulting in an overestimation of safety.

Virus Assays to Be Used

Whenever possible, virus assays should be infectivity based. There should be an efficient, sensitive, and reliable infectivity assay for the viruses used. Methods such as PCR or the RT assay can indicate the presence of virus and/or viral nucleic acid but cannot distinguish between infectious and noninfectious virus or viral components.

Model Viruses Used in Process Validation

Unlike bacterial validations where *Brevundimonas (Pseudomonas) diminuta,* ATCC 19146, serves as the industry-accepted standard for the validation of sterile filtration processes, there is no one viral agent that can, currently, serve as a generic model virus. The general approach to validation studies has been, therefore, to validate with a panel of viruses that includes known contaminants, suspected contaminants, or model viruses resembling suspected contaminants and a range of viruses of differing properties that are not themselves considered likely contaminants but are useful in assessing the rigor of the process. Table 6.8 lists the kinds of viruses that may be employed in validation studies.

In a discussion of viruses to be used in process validation studies, several terminologies require definition.

- *Relevant virus:* A virus that is of the same species as the viruses that are known, or likely, to be present as contaminants. However, use of a "relevant" virus may pose logistic problems, in that it may be too hazardous to work with or, alternatively, cannot be grown in vitro to high titers. For example, hepatitis B virus and hepatitis C virus are not easily propagated in cell culture. In such instances, a model virus is used. It is important to note that even when using a relevant virus (such as HIV in the case of blood products), it could always be argued that the relevant virus is, in fact, a laboratory strain, grown in vitro, and may differ from the

Table 6.8. Kinds of Viruses Used in Virus Validation Studies

Type of Virus	Test Virus	Reference
Relevant virus	HIV	ICH (1997)
Specific model virus	• Murine leukemia virus as a model for endogenous retroviruses	ICH (1997)
	• Duck hepatitis B virus as a model for hepatitis B virus	Long et al. (1993)
	• BVDV as a model for hepatitis C virus	Francki et al. (1992)
Nonspecific model virus	• Poliovirus (as a representative of a small virus)	ICH (1997)
	• SV-40	ICH (1997)
	• Parvovirus, canine/ porcine (as a model for parvovirus B-19)	CPMP (1991)
	• HSV (as a representative of a large DNA virus)	ICH (1997)
	• Pseudorabies virus (as a representative of herpes viruses)	ICH (1997)
Surrogate virus	• Bacteriophage φ6 as a model for viruses in the 80–120 nm size range	Aranha-Creado et al. (1998); Lytle et al. (1992)
	• Bacteriophage PR772 as a model for viruses larger than 50 nm	Aranha-Creado et al. (1997)
	• Bacteriophages PP7, φX-174 as models for viruses in the 25–30 nm size range	Oshima et al. (1995b, 1996; DiLeo et al. (1993a, b)

naturally occurring strains of viruses of differing properties that are not themselves considered likely contaminants.

- *Specific model virus:* In the event that process validation cannot be undertaken with a relevant virus (which is very often the case), using a specific model virus is recommended. A specific model virus is one that is closely related to the known or suspected virus (same genus or family), having similar physical and chemical characteristics as the observed or suspected virus. For example, a murine retrovirus such as murine leukemia virus is an appropriate model virus when cells of murine origin are used because of the potential of contamination of these cell lines with endogenous retroviral particles. Similarly, viruses such as hepatitis B and C are well characterized in molecular and virological terms but can only be grown and assayed for infectivity in humans or primates. Suggested models for hepatitis C include BVDV, yellow fever virus, or sindbis virus (Francki et al. 1992); duck hepatitis virus has been used as a model for hepatitis B (Long et al. 1993).

- *Nonspecific model virus:* In addition to demonstrating removal of known or potential viral contaminants, the validation study is also required to demonstrate the ability of the manufacturing process to clear putative viruses or unknown viral variants. Thus, nonspecific model viruses are included to evaluate the robustness of the clearance process for a wide range of viruses. Generally, the panel of viruses includes a representative selection of DNA and RNA viruses, nonenveloped and enveloped viruses, small and large viruses, and resistant and nonresistant viruses. Nonspecific model viruses that have been included are polio, SV-40 or an animal parvovirus as small nonenveloped viruses, a parainfluenza or a murine retrovirus as large enveloped RNA virus, and a herpes virus as a large DNA virus. These viruses are unlikely to be present in the source material but are useful in assessing the rigor of the process (CPMP 1996; ICH 1997).

- *Surrogate viruses:* The use of surrogates for mammalian viruses, e.g., bacteriophages, is justified when the method of removal is dependent on size exclusion, i.e., filtration (Aranha-Creado and Brandwein 1998), and not a particular criterion/constituent associated with the virus, as is the case with solvent-detergent inactivation (presence of a lipid envelope) or partitioning (where specific physicochemical properties of the virus influence its interaction with gel matrices and precipitation properties). Bacteriophages PP7 and ϕX-174 have been recommended as surrogates for poliovirus in studies evaluating virus removal by filtration (DiLeo et al. 1993a, b; Oshima et al. 1995b; Oshima et al. 1996). Similarly, ϕ6 has been used as a surrogate for murine leukemia virus and HIV to evaluate viral clearance of endogenous retroviruses (Aranha-Creado et al. 1998) and blood-borne viruses (Lytle et al. 1992). Tables 6.9 and 6.10 list some of the viruses that have been used in validation studies.

Calculation of Virus Reduction Factor

The virus reduction factor (VRF) of an individual purification or removal/inactivation step is defined as the $\log_{10}$ of the ratio of the virus load in the prepurification material divided by the virus load in the post-purification material:

$$\text{VRF} = \log_{10} \frac{\text{virus load in the prepurification material}}{\text{virus load in the postpurification material}}$$

A clearance factor for each stage can be calculated, and the overall clearance capacity of the production process assessed.

The total VRF for the entire manufacturing operation is calculated as the sum of the log reduction factors for individual purification steps, provided that the various steps possess fundamentally different mechanisms of viral clearance:

$$\text{Total VRF} = \text{VRF}_1 + \text{VRF}_2 + \ldots \text{VRF}_n$$

Reductions in viral titer of the order of one log or less are considered unreliable because of the limitations of viral validation studies and are not to be included in the calculation of reduction factors (ICH 1997). Additionally, since viruses vary greatly with regard to their inactivation or removal profiles, only data for the same model virus can be cumulative. Guidelines specifically state that confidence intervals should be calculated for all studies with model viruses, and the confidence intervals for both the preprocessing titer and the postprocessing titer should be included in the confidence interval for the process reduction factor. Specifically, the confidence interval for reduction factor calculations should be computed with confidence intervals equal to $\pm S1 + A1$, where S is the 95 percent confidence interval for the preprocess material, and A is the 95 percent confidence interval for the postprocess material.

Table 6.9. Examples of Mammalian Viruses That Have Been Used in Validation Studies

Virus	Family (-viridae)	Genome	Envelope	Size (nm)	Shape
Pseudorabies virus	Herpes	DNA	Yes	120–300	Spherical
HSV	Herpes	DNA	Yes	120–300	Spherical
HIV	Retro	RNA	Yes	80–100	Spherical
Murine leukemia virus	Retro	RNA	Yes	80–120	Spherical
Reovirus 3	Reo	RNA	No	60–80	Spherical
Sindbis virus	Toga	RNA	Yes	40–70	Spherical
SV-40	Papova	DNA	No	40–55	Icosahedral
BVDV	Toga	RNA	Yes	40–70	Pleomorphic/ Spherical
Encephalomyo-carditis virus	Picorna	RNA	No	22–30	Icosahedral
Poliovirus	Picorna	RNA	No	25–30	Icosahedral
Hepatitis A	Picorna	RNA	No	25–30	Icosahedral
Parvovirus (canine, porcine)	Parvo	DNA	No	18–26	Icosahedral

Table 6.10. Examples of Bacteriophages That Have Been Used in Validation Studies

Virus	Family (-viridae)	Host	Genome	Envelope	Size (nm)	Shape
φ6	Cysto	*Pseudomonas phaseolicoli*	RNA	Yes	75–86	Spherical
PR772	Tecti	*Escherichia coli* K12	DNA	No	53	Icosahedral
PP7	Levi	*Pseudomonas aeruginosa*	RNA	No	25	Icosahedral
MS-2	Levi	*Escherichia coli* K12	RNA	No	25	Icosahedral
φX174	Micro	*Escherichia coli* C	DNA	No	25–27	Icosahedral

Validation of Viral Clearance

The above section delineated the complexity of demonstrating and documenting viral clearance. It must, therefore, be recognized that virus validation studies are only an approximation of the true capacity of a process in view of the complex variables involved.

- It is almost impossible to reproduce exactly the full manufacturing conditions in the laboratory and, thus, slight differences may occur between the scaled-down (laboratory) process and that of the full manufacturing scale.

- Model viruses may not behave in the same manner as wild type viral contaminants, leading to a degree of subjective error in the calculation of virus reduction factors.

- Even though every attempt is made to include an appropriate panel of viruses in validation, virological relatedness is not necessarily able to predict accurately the responses of related viruses to specific environmental parameters; for example, both poliovirus and hepatitis A virus are picorna viruses, but they differ considerably in their resistance to physicochemical inactivation methods. Additionally, in natural populations, there may exist fractions that are more resistant than others to inactivation.

- Errors may also be made in calculating the overall clearance capacity of the production process. Adding together the removal of a virus by a series of steps may give an artificially high value, especially if the steps are very similar to each other (e.g., multiple steps with similar types of buffers, similar columns used for chromatography, etc.) Therefore, it is recommended that at least one of the stages in the production process should be able to remove or inactivate five logs or more of virus.

It is clear that studies can only approximate the real situation. Experimental limitations (e.g., laboratory virus strain versus wild type strain, summing up of logarithmic reduction factors for several steps) should always be kept in mind. The procedures used should reflect efforts to come to a conclusive risk assessment.

METHODOLOGIES FOR VIRUS
REMOVAL BY FILTRATION

In general, filtration processes operate via one of two mechanisms: size exclusion or adsorptive retention. Size exclusion is the mechanism of choice due to geometric or spatial restraint. Removal by size exclusion is not directly influenced by filtration conditions (e.g., differential pressure, temperature, viral challenge level, etc.) and characteristics of the product to be filtered (e.g., viscosity, ionic strength, pH, surface tension, etc.); therefore, it provides a predictable mechanism of viral particle removal. In the case of adsorptive retention, other mechanisms can also influence removal: for example, adsorption to the filter by electrokinetic or hydrophobic interactions and filtration conditions (flow rate, pressure, and temperature). Factors that could potentially affect retention by adsorption include filter type (structure, base polymer, surface modification chemistry, pore size distribution, thickness), fluid components (formulation, surfactants, additives), fluid properties (pH, viscosity, osmolarity, ionic strength), and process conditions (temperature, pressure differential, flow rate, time). While both these mechanisms operate concomitant with each other, the relative importance and role of each may vary.

Membrane filter efficiency is evaluated in terms of the titer reduction or log titer reduction (LTR). Titer reduction is the ratio of the particle (bacterial/viral) concentration per unit volume in the prefiltration suspension to the concentration per unit volume in the postfiltration suspension. Similarly, LTR is the $\log_{10}$ of this ratio. It should be noted that in the case of sterilizing grade filters for bacterial removal, the bacterial retention requirement is clearly defined and has been extensively validated in manufacturing environments in the last few decades. By pharmaceutical industry agreement and FDA definition, sterilizing grade filters, typically rated at 0.2 μm, are filters that can remove more than 10^7 colony forming units (CFU)/cm^2 of *Brevundimonas diminuta* and yield a sterile effluent. In the case of virus removal filters, as discussed in earlier sections, the unique considerations associated with the documentation of viral safety currently preclude stipulation of a particular test virus for validation or specification of the exact viral load to be used in validation experiments.

The two major membrane filtration systems for the removal of viral particles from fluids are single pass or direct flow filtration and cross-flow or tangential flow filtration. The following sections briefly describe applications of filtration for virus removal.

Direct Flow Filtration

Conventionally, membranes used for microfiltration are symmetric microporous structures with pore sizes in the range of 0.1 to 10 μm (Santos et al. 1991). In the single pass or direct flow filtration mode, the entire volume is passed through the membrane filter. The flow of the liquid is perpendicular to the filter surface and the particles associated with the liquid being filtered are deposited either on or within the membrane (Figure 6.1). Eventual clogging of the filter pores is therefore an inevitable consequence. Typically, direct flow filters are designed to be used once. The advantages of direct flow microfiltration include ease and speed of use, low shear, and high levels of product recovery.

Various membrane filters with virus retention claims are currently available. A polyvinylidene fluoride (PVDF) membrane filter (Ultipor® VF grade DV50 virus removal filter, Pall Corporation, East Hills, N.Y.) has been shown to provide ≥ 6 log removal for viruses > 50 nm in size, independent of fluid type (Aranha-Creado et al. 1997; Oshima et al. 1996; Roberts 1997). Virus removal is effected predominantly by a size exclusion mechanism, and the retention of viruses larger than 50 nm is independent of the type of virus (i.e., enveloped/nonenveloped, DNA/RNA, etc.) or process fluid (Table 6.11). Removal of viruses smaller than 50 nm may occur in a consistent process fluid specific manner (Aranha-Creado et al. 1997; Oshima et al. 1996). Minimum levels of virus removal can be established for these systems if fluid and process conditions are used that

Figure 6.1. Flow dynamics in direct flow filtration.

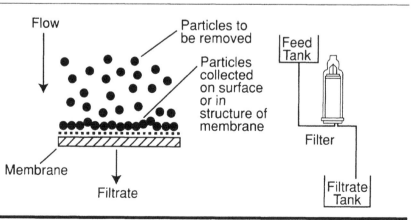

Table 6.11. Removal of Viruses from Fluids by Direct Flow Filtration

Filter Type	Virus[a]	Fluid[b]	LTR	Reference
PVDF[c]: Ultipor VF grade DV50	Influenza A virus	MEM + 10	> 6.3	Aranha-Creado et al. (1997)
	Poliovirus	MEM + 10	2.2	
		Water	4.5	Roberts (1997)
	Vaccinia virus	PBS + 5% HSA	> 6.4	
	HSV-1	PBS + 5% HSA	> 5.8	
	Sindbis	PBS + 5% HSA	> 5.6	
	Semliki forest virus (SFV)	PBS + 5% HSA	> 8.5	
	Poliovirus	PBS + 5% HSA	< 1	
Nylon 66; Ultipor N66 grade NDP	Influenza A virus	MEM + 10	> 6.6	Oshima et al. (1994)
Regenerated cellulose hollow fiber; Planova 35N	HIV	Blood coagulation factor IX	7.8	Burnouf-Radosevich et al. (1994)
	BVDV		> 5.9	
	Reovirus type 3		> 6.1	
	BPV		> 5.8	
	SV-40		> 7.8	
Planova 15N	Poliovirus		> 6.7	Burnouf-Radosevich et al. (1994)
	BPV		> 6.3	

[a]Sizes of viruses are as follows: BPV, 18–26 nm; BVDV, 40–70 nm; HSV-1, 120–300 nm; HIV, 80–100 nm; Influenza A virus, 80–120 nm; Poliovirus, 28–30 nm; Reovirus type 3, 60–80 nm; SFV, 40–70 nm; SV-40, 40–55 nm; Sindbis, 40–70 nm; *Vaccinia* virus, 250–260 × 300–450 nm.

[b]Carrier fluids as follows: MEM + 10, Dulbecco's modified Eagle medium supplemented with 10% fetal bovine serum; PBS, phosphate-buffered saline; HSA, human serum albumin

[c]Polyvinylidene fluoride

minimize removal of viral particles by mechanisms other than size selection.

A cuprammonium-regenerated cellulose hollow fiber microfilter has been used for viral clearance from coagulation factor concentrates and other biologicals (Hamamoto et al. 1989; Sekiguchi et al. 1989). Using the Planova 15N and Planova 35N (Asahi Chemical Industry Co., Ltd., Tokyo, Japan) in the single direct flow filtration mode, Burnouf-Radosevich et al. (1994) reported several logs of clearance for large and small viruses (Table 6.11). Bovine parvovirus (BPV), a relatively small virus (~25 nm in size) was removed, just as efficiently, by both the Planova 15N and 35N filters. It must be noted that while the authors attributed removal of a virus smaller than the pore size of the membrane to aggregation effects, they did not clarify whether BPV aggregates were present in the BPV stock (and, therefore, an artifact of the validation study) or whether it existed in a naturally aggregated state in their product.

Using Nylon 66 filter membranes (Ultipor® Nylon 66 Grade NDP filters, Pall Corporation, East Hills, N.Y.), Oshima and coworkers (1994; 1995a) demonstrated removal of influenza A virus (80–120 nm) and HIV (80–100 nm) to below detectable limits in all fluids tested. However, removal of viruses smaller than influenza A virus was not as efficient. Titer reduction results for small viruses, i.e., 25–50 nm in size, varies depending on fluid type. The highest titer reductions were observed in carrier solutions such as water and solutions with low concentrations of protein while smaller titer reductions were observed in solutions containing serum. These results suggest that other factors in addition to size exclusion were enhancing the titer reduction for viruses in the 25–50 nm size range. Other researchers have reported diminished adsorptive effects in the presence of serum or pretreating a normally adsorbent membrane with serum or gelatin (Cliver 1965).

Tangential Flow Filtration

In the cross-flow or tangential flow filtration (TFF) mode, the liquid containing a mixture of different size components is brought to the surface of a semipermeable membrane; a portion of the liquid passes through the membrane surface (permeate), while the rest is returned to the central reservoir as the retentate (Figure 6.2). The liquid flow on the upstream side of the filter is tangential to or across the filter surface; this results in a liquid sweep of the membrane surface and avoids progressive buildup of particles. In this process, the volume of fluid in the retentate continually decreases

Figure 6.2. Flow dynamics in cross-flow filtration.

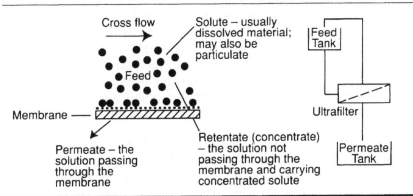

as more of the initial volume is collected as permeate. Viral particles are concentrated in the retentate. Typically, ultrafiltration systems utilizing cross-flow are either in the tangential flow configuration, where fluid passes between two flat sheets of membrane material, or hollow fiber filters, where the fluid passes through the middle of hollow tubes.

Ultrafiltration membranes are asymmetric or "anisotropic"; they are composed of an ultrathin "skin" layer (thickness of 1 ϕm or less) supported on a highly porous sublayer (100–200 ϕm in thickness). The "skin" layer is responsible for the exclusion characteristics of the membrane; the sublayer supports the thin layer and contributes to the high transmembrane fluxes displayed by asymmetric membranes compared with symmetric membranes of the same thickness. Asymmetric membranes are less easily plugged as the particles above the nominal molecular weight cutoff of the membrane are retained at the surface; thus, cleaning of these membranes primarily involves the removal of surface deposition. In contrast, in the case of microporous membranes, particle deposition occurs not only at the surface; penetration and clogging of the pores also occurs. It must be noted, however, that while ultrafilters have been traditionally designed for reuse after cleaning and sanitization, reuse is currently not the norm for virus removal applications because of the complexity of the validation issues involved.

A composite PVDF membrane has been reported to retain viruses predominantly by a sieving mechanism (DiLeo et al. 1993a, b). Using the Viresolve/70™ membrane (Millipore Corporation, Bedford, Mass.), DiLeo et al. demonstrated retention of model mammalian viruses, ranging from 3.5 logs with poliovirus to greater than 6.8 logs with murine leukemia virus. Viral clearance was augmented in the presence of serum. This is to be expected due to the accumulation of proteinaceous materials on the filter surface. This phenomenon is commonly referred to as gel polarization and essentially refers to the accumulation at the membrane surface of solutes rejected by the membrane. While the gel layer enhances the retentivity of the membrane, excessive gel polarization may result in drastically reduced flow rates and eventual fouling of the membrane.

Modified polyethersulfone (PES) membranes, the Pall Filtron® Omega™ 300 K VR and 100 K VR (Pall Filtron Corporation, Northborough, Mass.) have been reported to retain viruses by a size exclusion–based mechanism. Aranha-Creado and Herczeg (1997) and others (Pall Corp. 1998) have reported retention of greater than 4 logs of murine leukemia virus and porcine parvovirus by the Omega™ 300 K VR and 100 K VR membranes, respectively (Table 6.12).

The virus retention ability of hollow fiber ultrafilters composed of either polyacrylonitrile (PAN) or polysulfone (PS) (Microza®, Pall Corporation, East Hills, N.Y.) has been studied by Oshima et al. (1995b). These investigators reported a greater than 6.5 LTR for poliovirus by PES with a 13,000 molecular weight cutoff and a 6,000 molecular weight cutoff for PAN hollow fiber ultrafilters in a medium supplemented with serum (Table 6.12).

INTEGRITY TESTING

From a manufacturing standpoint, in addition to demonstrating consistent viral clearance concomitant with high product recovery, a filter user must be able to document the filter performance characteristics under process conditions. While a particulate (bacterial/viral) challenge constitutes the only true test of a filter's ability to retain these biological species, it is a destructive test that precludes subsequent use of the filter. Additionally, introduction of viruses into the production environment is against GMPs. Therefore, filter manufacturers provide validation documentation for a filter, which includes correlation of integrity test values with retention of an appropriate test bacteria or virus. For example, in the case of the Ultipor® VF grade DV50 filter, the manufacturer has provided a nondestructive forward flow test correlated to virus removal. The test involves wetting the filter, applying gas pressure (typically air or nitrogen), and measuring the resultant transmembrane gas flow

Table 6.12. Removal of Viruses by Tangential Flow Filtration

Filter Type[a]	Virus[b]	Fluid[c]	LTR	Reference
PES; Omega™ 300 K VR	Murine leukemia virus	PBS	> 4.0	Pall Corp. 1998
Omega™ 100 K VR	Porcine parvovirus	PBS	> 4.0	Aranha-Creado and Herczeg (1997)
PVDF; Viresolve/70™	Polio	PBS	3.1–3.51	DiLeo et al. (1993a)
		PBS + HSA	4.2	
	SV-40	PBS	4.9–5.7	
		PBS + HSA	> 5.7	
	Sindbis	PBS	7.4	
	Reovirus-3	PBS	7.2	
	Reovirus -3	PBS + HSA	> 7.6	
PAN, hollow fiber, Microza®; 50 K 13 K	Polio Polio	MEM + 10 MEM + 10	4.6 > 6.5	Oshima et al. (1995)
PS, hollow fiber, Microza®; 6 K	Polio	MEM + 10	>6.4	Oshima et al. 1995

[a]PES, polyethersulfone; PVDF, polyvinylidene fluoride; PAN, polyacrylonitrile; PS, polysulfone.
[b]Sizes of viruses are as follows: murine leukemia virus, 80–120 nm; poliovirus, 28–30 nm; porcine parvovirus, 18–26 nm; reovirus, type 3, 60–80 nm; Simian virus 40 (SV-40), 40–55 nm; Sindbis, 40–70 nm.
[c]Carrier fluids as follows: MEM + 10, Dulbecco's modified Eagle medium supplemented with 10% fetal bovine serum; PBS, phosphate-buffered saline; HSA, human serum albumin.

across the wetted membrane. A filter that tests at or below the limit forward flow value of 12.5 cc/min per 10 in. element (set by the filter manufacturer) will assuredly provide an LTR of 6 or greater for any virus larger than 50 nm (Pall Corp. 1995). Figure 6.3 gives the validation data for the grade DV50 elements (Aranha-Creado et al. 1997a). This approach of correlation of integrity test values (provided by the filter manufacturer) with virus removal provides the filter user with an assurance of safety without compromising filter utility in production. Ideally, under manufacturing conditions, the filter assembly is subjected to a nondestructive integrity test before and after completion of the filtration operation; however, minimally, a postuse integrity test is mandated following filtration to document the performance characteristics of the filter.

In the case of sterile filtration applications for bacterial removal, the industry-accepted integrity tests are similarly performed by applying gas (air or nitrogen) pressure to a wetted filter and monitoring the transmembrane airflow. The types of tests include the bubble point test, the forward flow test, and the pressure hold test (Brantley and Martin 1997). The bubble point is dependent on

Figure 6.3. Correlation of forward flow limit (isopropyl alcohol/water, 30:70 wet at 85 lb/in.^{2}) with log titer reduction (for the model virus, bacteriophage PR772) for Ultipor® VF™ AB1UDV50 filter elements (Aranha-Creado et al. 1997a)

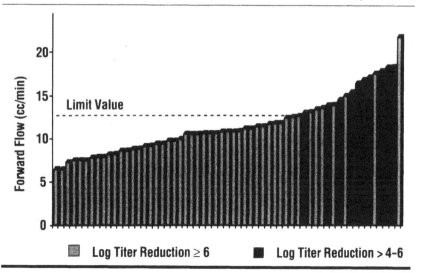

observation downstream of bubbling, which is indicative of bulk flow, through the filter, resulting from wetting fluid being expelled from the pores, followed by air bubbles (Figure 6.4). High bubble point pressures are associated with finer grades (smaller pores) of filter medium. In the case of virus retentive filters, with pore size in the nanometer size range, the pressure required to observe the bubble point for these membranes is in excess of 300 psi when wet with water or aqueous solutions with comparable surface tension.

The forward flow test quantitatively measures the diffusive flow of gas as well as flow through any open pores in a wetted membrane filter (Figure 6.5). The test is performed by wetting a membrane filter and applying a predetermined constant pressure. The test pressure (established for a particular filter by the filter manufacturer) is typically less than the bubble point for the filter. The test pressure is set at a pressure where the diffusional flow is stable and low, so as not to obscure bulk flow. The diffusional gas (air or nitrogen) flow rate, as well as the flow through any open pores, is measured through the wetted membrane. The gas flow is

Figure 6.4. Schematic representation of the bubble point test. The bubble point is the pressure at which bulk flow is initiated through the largest pore in the membrane. The filter is wet with the test fluid; air (or nitrogen) pressure is applied and slowly increased until the test fluid is expelled from the largest pore in the filter, at which time vigorous bubbling of air is observed in the downstream collection vessel.

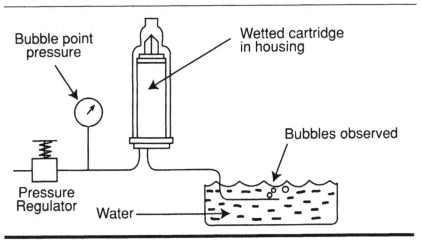

Figure 6.5. Schematic representation of the forward flow test. The diffusive or forward flow test quantitatively measures the gas flow through all pores of the wetted filter. Diffusional flow and flow through the open pores is measured on the downstream side.

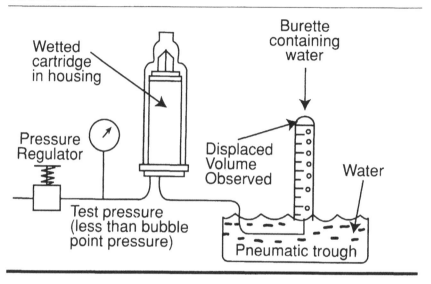

defined as a volume per specified time period (mL/min). The filter is integral if the forward flow value is lower than the manufacturer's specified value.

The pressure hold test is a modified form of upstream forward flow testing and involves the measurement of decay in pressure over a specified time period for a given filter assembly and wetting fluid (Figure 6.6). Since the pressure hold test is performed upstream (and, consequently, the downstream sterile connections are not disturbed), the advantage of this test is that it can be performed after sterilization of the filter assembly and before *and* after filtration. Essentially, the filter housing is pressurized to the test pressure specified by the filter manufacturer; then the filter is isolated from the pressure source. The diffusion of gas across the wetted membrane is measured as a decay in pressure over a specified period of time. The pressure hold and forward flow tests are related through the ideal gas law.

Filters in virus removal applications must be at least integrity tested following filtration to ensure the performance of the filter.

Figure 6.6. Schematic representation of the pressure hold test. The diffusive or forward flow test quantitatively measures the gas flow through all the pores of the wetted filter. Pressure decay is observed on the pressure gauge, located on the upstream side, during the specified test time.

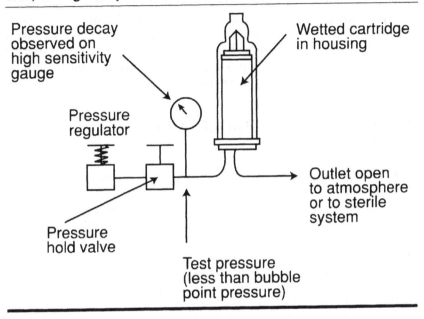

Ideally, the integrity test should be performed both before and after use; this is possible when a nondestructive integrity test method is used. The manufacturer-recommended forward flow test for the Ultipor® VF™ grade DV50 virus removal filter should be conducted in either 30 percent isopropyl alcohol or 20 percent ethanol (Pall Corp. 1995). This test, being nondestructive, is amenable to use both pre- and postfiltration and is correlated to virus removal by the filter manufacturer.

Another nondestructive test is the liquid porosimetric integrity test, which is also correlated to virus removal. The manufacturer-recommended integrity test for the Viresolve modules is the Corrtest™ (Phillips 1996), which is a liquid-liquid intrusion integrity test; the air-liquid interface typically encountered in bubble point/diffusion tests is replaced with the interface between two immiscible liquids. It is essentially a ratio of two membrane permeabilities measured at preselected operating conditions using a pair of mutually immiscible fluids, one of which is employed as a membrane wetting agent and the other as an intrusion fluid.

A gold particle removability test is the manufacturer-specified integrity test for the Asahi Planova filters (Sato et al. 1994). It involves challenge of the filter with a colloidal gold suspension followed by determination of the concentration of these particles in the pre- and postchallenge material by spectrophotometric methods and calculation of the log rejection coefficient. This test is destructive and, therefore, is applicable only for postuse testing. In general, the integrity test results must correlate with the virus removal claims as specified by the filter manufacturer.

POSITIONING OF VIRAL CLEARANCE METHODOLOGIES

The current approach to viral containment (as indicated in Table 6.7) is aimed at the prevention of entry of viral contaminants (through adequate sourcing and placement of appropriate barriers to viral entry) as well as incorporation of viral clearance methodologies. The positioning of the viral containment methodology will depend on several factors, including, but not limited to, the kind of process involved; the nature of the viral contaminant, i.e., actual (endogenous/adventitious virus) or theoretical; whether the method is based on virus inactivation or removal; and whether the particular strategy is based on serendipitous clearance or is a deliberate method incorporated for virus clearance.

When the virus reduction barrier is positioned upstream of the fermentors for the removal or inactivation of adventitious viruses from raw material streams, it is generally referred to as a virus barrier (Figure 6.7). Examples of virus barriers are inactivation (e.g., heat, irradiation, chemical) and removal (e.g., filtration) methods that may be incorporated to reduce the viral burden of raw materials, such as media constituents and fermentor air supply, that are introduced into the fermentor. Inactivation methods, such as heat/radiation/chemical treatments, may alter the biological integrity of media formulations/heat labile growth supplements and may affect the medium's ability to support growth in fermentation processes. Filtration as a viral barrier is the least intrusive; it should have no effect on cell growth and target protein production and should be readily amenable to use in currently licensed processes. In general, when the virus reduction methodology is incorporated downstream of the fermentor, it is referred to as a viral clearance method.

Figure 6.7. Possible positioning of virus reduction methodologies.

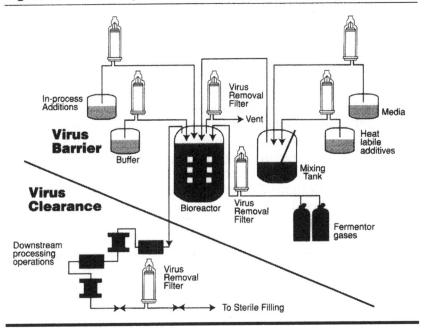

Filtration processes for viral clearance may be incorporated in the latter stages of purification processes to ensure exclusion of endogenous or adventitious contaminants that may gain access to the product through manufacturing raw materials or during the manufacturing process. When thus positioned, it ensures improved filter life (in view of the decreased particulate load) and also provides a significant level of assurance that the product will be free of adventitious agents. Currently, both ultrafiltration and microfiltration are well established steps for virus removal from biological solutions; filtration, for virus reduction, has been used in both the direct flow filtration mode (Aranha-Creado et al. 1997; Hamamoto et al. 1989; Oshima et al. 1996; Roberts 1997; Sekiguchi et al. 1989), and TFF mode (DiLeo et al. 1993a, b). Additionally, a combination of tangential flow and direct flow methodologies (Aranha-Creado and Fennington 1997) can serve as multiple orthogonal approaches to effect necessary viral clearance, while at the same time providing benefits from a logistic and economical standpoint.

CONCLUSIONS

Recognition of the potential for viral contamination of biologicals by real and putative viruses has resulted in a concerted approach, during manufacture, involving rigorous and effective process controls and the incorporation of multiple barriers to virus dissemination so as to provide overlapping and complementary levels of protection. Several virus removal and inactivation methods are currently available, and their judicious use will enhance the level of safety associated with these products. Filtration, especially when based on size exclusion, is a reliable and "robust" method of viral clearance. The ease of use of filtration systems, along with the inherent nondestructive nature of the filtration process, makes it readily amenable in process applications. Currently available filtration devices (which operate in either the direct flow or tangential flow mode) provide high product transmission along with consistent viral clearance. Filtration methodologies designed for viral removal applications will contribute significantly to the reduction of endogenous and adventitious viral contaminants and concomitantly will enhance the virological safety of biologicals and biopharmaceuticals.

REFERENCES

Aach, R. D., and R. A. Kahn. 1980. Post-transfusion hepatitis: Current perspectives. *Ann. Intern. Med.* 92:539–546.

Alter, H. J., Y. Nakatsuji, J. Melpolder, J. Wages, R. Wesley, W. Shih, and J. P. Kim. 1997. The incidence of transfusion-associated hepatitis G virus infection and its relation to liver disease. *N. Engl. J. Med.* 336:747–754.

Anderson, K. P., Y. S. Lie, M. L. Low, S. R. Williams, F. M. Wurm, and Dinowitz, M. 1991. Defective endogenous retrovirus-like sequences and particles of Chinese hamster ovary cells. *Dev. Biol. Stand.* 75:123–132.

Aranha-Creado, H., and H. Brandwein. 1998. Application of bacteriophages as surrogates for mammalian viruses: A case for use in filter validation based on precedents and current practices in medical and environmental virology. *Scientific and Technical Report,* PBB-STR 39. East Hills, N.Y., USA: Pall BioPharmaceuticals, pp. 1–15.

Aranha-Creado, H., and G. J. Fennington. 1997. Cumulative viral titer reduction demonstrated by sequential challenge of a tangential flow membrane filtration system—Omega™ 300 K VR Maximate™ system—and a direct flow pleated filter cartridge—the Ultipor® VF Grade DV50 virus removal filter. *PDA J. Pharm. Sci. Technol.* 51:208–212.

Aranha-Creado, H., and H. Herczeg. 1997. Demonstration of parvovirus clearance by a tangential flow membrane filtration system—Pall Filtron Omega™ 100 K VR. *Scientific and Technical Report*, STR-FIL 1. East Hills, N.Y., USA: Pall Ultrafine Filtration Co., pp. 1–8.

Aranha-Creado, H., J. Peterson, and P. Y. Huang. 1998. Clearance of murine leukemia virus from monoclonal antibody solutions by a hydrophilic PVDF microporous membrane filter. *Biologicals* 26 (May/June).

Aranha-Creado, H., K. Oshima, S. Jafari, G. Howard, Jr., and H. Brandwein. 1997. Virus retention by a hydrophilic triple-layer PVDF microporous membrane filter. *J. Pharm. Sci. Technol.* 51:119–124.

Barrowcliffe, T. W. 1993. Viral inactivation vs. biological activity. *Dev. Biol. Stand.* 81:125–135.

Bergmann, D. G., and D. A. Wolff. 1981. Production of R-type virus-like particles in hamster cells. *Intervirol.* 16:61–70.

Blackbourn, D. J., J. Ambroziak, E. Lennette, M. Adams, B. Ramachandran, and J. A. Levy. 1997. Infectious human herpes virus 8 in a healthy North American blood donor. *Lancet* 349: 609–611.

Bowen, G. S., C. H. Calisher, W. G. Winkler, A. L. Kraus, E. H. Fowler, R. H. Garman, D. W. Fraser, and A. R. Hinman. 1975. Laboratory studies of a lymphocytic choriomeningitis virus outbreak in man and laboratory animals. *Am. J. Epidemiol.* 102:233–240.

Brantley, J. D., and J. M. Martin. 1997. Integrity testing of sterilizing grade filters. *Scientific and Technical Report*, PBB-STR-28. East Hills, N.Y., USA: Pall Ultrafine Filtration Co., pp. 1–8.

Brown, K. E., and N. S. Young. 1997. Parvovirus B19 in human disease. *Annu. Rev. Med.* 48:59–67.

Brown, P., M. A. Preece, and R. G. Will. 1992. "Friendly fire" in medicine: Hormones, homografts, and Creutzfeldt-Jakob Disease. *Lancet* 340 (8810):24–27.

Builder, S., R. Van Reis, N. Paoni, and J. Ogez. 1989. Process development and regulatory approval of lisane-type plasminogen activator. In *Advances in animal cell biology and technology for bioprocesses,* edited by R. W. Spier, J. B. Griffiths, J. Stephenne, and P. J. Crooy. Stoneham, Mass., USA: Butterworths, pp. 452–464.

Buller, R. M. L., A. Corman-Weinblatt, A. W. Hamburger, and G. D. Wallace. 1987. Observations on the replication of ectromelia virus in mouse-derived cell lines: Implications for epidemiology mousepox. *Lab. Animal Sci.* 37:28–32.

Burnouf, T. 1993. Chromatographic removal of viruses from plasma derivatives. *Dev. Biol. Stand.* 81:199–209.

Burnouf-Radosevich, M., P. Appourchaux, J. J. Huart, and T. Burnouf. 1994. Nanofiltration, a new specific virus elimination method applied to high-purity factor IX and factor XI concentrates. *Vox Sang.* 67:132–138.

Carthew, P. 1989. Possible significance of rodent virus contamination of biological products for use in humans. *Dev. Biol. Stand.* 70:133–134.

Centers for Disease Control. 1996. Hepatatis A among persons who received clotting factor concentrate–United States, September–December 1995. *Morb. Mortal. Weekly Rep.* 45 (2):29–32.

Cliver, D. O. 1965. Ionic composition of the viral suspension has also been shown to be important in some filters. *Appl. Microbiol.* 13:417–425.

Collinge, J., K. C. L. Sidle, J. Meads, J. Ironside, and A. F. Hill. 1996. Molecular analysis of prion strain variation and the aetiology of "new variant" CJD. *Nature* 383:685–690.

CPMP–Ad Hoc Working Party on Biotechnology/Pharmacy and Working Party on Safety Medicines. 1991. Note for guidance, validation of virus removal and inactivation procedures–EEC regulatory document. *Biologicals* 19:247–251.

CPMP. 1996. *Note for guidance on virus validations: The design, contribution and interpretation of studies validating the inactivation and removal of viruses.* London: European Agency for the Evaluation of Medicinal Products, Committee for Proprietary Medicinal Products; 29 February 1996.

CPMP. 1997. *Note for guidance on plasma-derived medicinal products.* London: European Agency for the Evaluation of Medicinal Products, Committee for Proprietary Medicinal Products; 27 June 1997.

Darby, S. C., D. W. Ewart, P. L. Giangrande, P. J. Dolin, R. J. D. Spooner, and C. R. Rizza. 1995. Mortality before and after HIV infection in the complete UK population of haemophiliacs. *Nature* 377:79–82.

Dichtelmuller, H., D. Rudnick, B. Breuer, and K. H. Ganshirt. 1993. Validation of virus inactivation and removal for the manufacturing procedure of two immunoglobulins and a 5% serum protein solution treated with β-propiolactone. *Biologicals* 21:259–268.

DiLeo, A. J., D. A. Vacante, and E. Deane. 1993a. Size exclusion removal of model mammalian viruses using a unique membrane system, Part I: Membrane qualification. *Biologicals* 21:275–286.

DiLeo, A. J., D. A. Vacante, and E. Deane. 1993b. Size exclusion removal of model mammalian viruses using a unique membrane system, Part II: Module qualification and process simulation. *Biologicals* 21:287–296.

Dorpema, J. W. 1988. A proposed European Pharmacopoeia monograph for continuous cell lines derived from drugs. *Dev. Biol. Stand.* 70:113–124.

Emanoil-Ravier, R., F. Hojman, M. Servenay, J. Lesser, A. Bernardi, and J. Peries. 1991. Biological and molecular studies of endogenous retrovirus-like genes in Chinese hamster cell lines. *Dev. Biol. Stand.* 75:113–122.

Erickson, G. A., S. R. Bolin, and J. G. Lundgraf. 1991. Viral contamination of fetal bovine serum used for tissue culture: Risks and concerns. *Dev. Biol. Stand.* 75:173–175.

FDA. 1987. *Guidelines for sterile drug products produced by aseptic processing.* Rockville, Md., USA: Food and Drug Administration, Center for Drug Evaluation and Research.

FDA. 1993. *Points to consider in the characterization of cell lines used to produce biologicals.* Rockville, Md., USA: Food and Drug Administration, Center for Biologics Evaluation and Research.

FDA. 1997. *Points to consider in the manufacture and testing of monoclonal antibody products for human use.* Rockville, Md., USA: Food and Drug Administration, Center for Biologics Evaluation and Research.

Fox, J. P., C. Manso, H. A. Penna, and M. Pare. 1942. Observation on the occurrence of icterus in Brazil following vaccination against yellow fever. *Am. J. Hyg.* 36:68–116.

Francki, R. I. B., C. M. Fauquet, D. L. Knudson, and F. Brown. 1992. Classification and nomenclature of viruses: Fifth report of the

International Committee on Taxonomy of viruses. *Archives of Virology* Sup 2:216–233.

Fraumeni, Jr., J. F., C. R. Stark, E. Gold, and M. L. Lepow. 1970. Simian virus 40 in polio vaccine: Follow-up of newborn recipients. *Science* 167:59–60.

Garnick, R. L. 1996. Experience with viral contamination in cell culture. *Dev. Biol. Stand.* 88:49–56.

Glaser, R. 1988. Concerns for using Epstein-Barr virus positive human B-lymphoblastoid cell lines for the production of human monoclonal antibodies. *Dev. Biol. Stand.* 70:131–132.

Gleeson, M., L. Herd, and C. Burns. 1990. Effect of heat inactivation of HIV on specific serum proteins and tumour markers. *Ann. Clin. Biochem.* 27:592–594.

Gomperts, E. D. 1986. Procedures for the inactivation of viruses in clotting factor concentrates. *Am. J. Hematol.* 23:295–305.

Grun, J. B., E. M. White, and A. F. Sito. 1992. Viral removal/inactivation by purification of biopharmaceuticals. *BioPharm* 5 (9): 22–30.

Hamamoto, Y., S. Harada, S. Kobayashi, K. Yamaguchi, H. Iijima, S. Manabe, T. Tsurumi, H. Aizawa, and N. Yamamoto. 1989. A novel method for removal of human immunodeficiency virus (HIV): Filtration with porous polymeric membranes. *Vox Sang.* 56:230–236.

Hamatainen, E., H. Suomela, and P. Ukkonen. 1992. Virus inactivation during intravenous immunoglobulin production. *Vox Sang.* 63:6–11.

Harris, R. J. C., R. M. Dougherty, P. M. Biggs, L. N. Payne, A. P. Goffe, and A. E. Churchill. 1966. Contaminant viruses in two live virus vaccines produced in chick cells. *J. Hyg.* (Cambridge) 64:1–6.

Hay, R. J. 1991. Operator-induced contamination in cell culture systems. *Dev. Biol. Stand.* 75:193–204.

Heimburger, N., H. Schwinn, P. Gratz, G. Luben, G. Kumpe, and B. Herchenhan. 1981. Factor VIII concentrate, highly purified and heated in solution. *Haemostasis* 10 (Suppl. 1):204.

Horaud, F. 1993. Viral safety of biologicals. *Dev. Biol. Stand.* 81:17–24.

Horowitz, B. 1989. Investigations into the application of tri(n-butyl) phosphate/detergent mixtures to blood derivatives. *Curr. Stud. Hematol. Blood Transfus.* 56:83–96.

Horowitz, B., and M. S. Horowitz. 1984. Human leukocyte alpha-interferon preparations: Laboratory characterization and evaluation of clinical safety. In *Interferon: Research, clinical applications and regulatory considerations,* edited by K. C. Zoon, P. P. Noguchi, and T. Y. Lie. New York: Elsevier Science Publishing Co., Inc., pp. 41–53.

Horowitz, B., M. E. Wiebe, A. Lippin, and M. H. Stryker. 1985. Inactivation of viruses in labile blood derivatives. I. Disruption of lipid-enveloped viruses by tri(n-butyl)phosphate detergent combinations. *Transfusion* 25:516–522.

ICH. 1997. *Viral safety evaluation of biotechnology products derived from cell lines of human or animal origin.* Harmonized tripartite guideline, Step 4 Adoption Recommendation. Geneva: International Federation of Pharmaceutical Manufacturers Association, pp. 1–27.

King, A. M., B. O. Underwood, D. McCahon, J. W. Newman, and F. Brown. 1981. Biochemical identification of viruses causing the 1981 outbreaks of foot and mouth disease in the U.K. *Nature* 293:479–480.

Lackritz, E. M., G. A. Satten, J. Aberle-Grasse, R. Y. Dodd, V. P. Raimondi, R. S. Janssen, W. F. Lewis, E. P. Notari, and L. R. Petersen. 1995. Estimated risk of transmission of the human immunodeficiency virus by screened blood in the United States. *New. Engl. J. Med.* 26:1721–1725.

Lawlor, E., S. Graham, F. Davidson, P. L. Yap, C. Cunningham, H. Daly, and I. J. Temperley. 1996. Hepatitis A transmission by Factor IX concentrates. *Vox Sang.* 71:126–128.

Lawrence, J. E. 1993. Affinity chromatography to remove viruses during preparation of plasma derivatives. *Dev. Biol. Stand.* 81:191–197.

LeDuc, J. W. 1987. Epidemiology of Hantaan and related viruses. *Lab Animal Sci.* 37:413–418.

Lever, A. M., A. D. Webster, D. Brown, and H. C. Thomas. 1984. Non-A, non-B hepatitis occurring in agammaglobulinemic patients after intravenous immunoglobulin. *Lancet* 2 (8411):1062–1064.

Liptrot, C., and K. Gull. 1991. Detection of viruses in recombinant cells by electron microscopy. In *Animal cell technology:*

Developments, processes and products, edited by R. E. Spier, J. B. Griffiths, and C. MacDonald. Oxford, UK: Butterworth-Heinemann Ltd., pp. 653–656.

Long, Z., C-S. Sun, E. M. White, B. Horowitz, and A. F. Sito. 1993. Hepatitis B viral clearance studies using duck virus model. *Dev. Biol. Stand.* 81:163–168.

Lueders, K. K. 1991. Genomic organization and expression of endogenous retrovirus-like elements in cultured rodent cells. *Biologicals* 19:1–7.

Lytle, D., S. C. Tondreau, W. Truscott, A. P. Budacz, R. K. Kuester, L. Venegas, R. E. Schmukler, and W. H. Cyr. 1992. Filtration sizes of human immunodeficiency virus type 1 and surrogate viruses used to test barrier materials. *Appl. Environ. Microbiol.* 58:747–749.

Magrath, D. I. 1991. Safety of vaccines produced in continuous cell lines. *Dev. Biol. Stand.* 75:17–20.

Mahy, B. W. J., C. Dykewicz, S. Fisher-Hoch, S. Ostroff, M. Tipple, and A. Sanchez. 1990. Virus zoonoses and their potential for contamination of cell cultures. *Dev. Biol. Stand.* 75:183–189.

Mannucci, P. M. 1992. Outbreak of hepatitis A among Italian patients with haemophilia. *Lancet* 339 (8796):819.

Mannucci, P. M., and M. Colombo. 1988. Virucidal treatment of factor concentrates. *Lancet* 2 (8614):782–785.

Manuelidis, L. 1994. The dimensions of Creutzfeldt-Jakob disease. *Transfusion* 34:915–928.

Manuelidis, L., T. Sklaviadis, and E. E. Manuelidis. 1987. Evidence suggesting that PrP is not the infectious agent in Creutzfeldt-Jakob Disease. *EMBO J.* 6:341–347.

Marcus-Sekura, C. J. 1992. Validation of removal of human retroviruses. *Dev. Biol. Stand.* 76:215–223.

Minor, P. 1991. Virological aspects of the quality control of human monoclonal antibodies. *Dev. Biol. Stand.* 75:233–236.

Morgenthaler, J. J., and A. Omar. 1993. Partitioning and inactivation of viruses during isolation of albumin and immunoglobulins by cold ethanol fractionation. *Dev. Biol. Stand.* 81:185–190.

Murphy, P., T. Nowak, S. M. Lemon, and J. Hilfenhaus. 1993. Inactivation of hepatitis A virus by heat treatment in aqueous solution. *J. Med. Virol.* 41:61–64.

Nathanson, N., and A. D. Langmuir. 1963. The Cutter incident: Poliomyelitis following formaldehyde-inactivated poliovirus vaccination in the United States during the spring of 1955. II. Relationship of poliomyelitis to Cutter vaccine. *Am. J. Epidemiol.* 142:109–140.

Ochs, H. D., S. H. Fischer, F. S. Virant, M. L. Lee, H. S. Kingdon, and R. J. Wedgwood. 1985. Non-A, non-B hepatitis and intravenous immunoglobulin. *Lancet* i:404–405.

Ono, M. 1988. Detection and elimination of endogenous retroviruses and retrovirus-like particles in continuous cell lines. *Dev. Biol. Stand.* 70:69–81.

Oshima, K. H., T. W. Comans, A. K. Highsmith, and E. A. Ades. 1995a. Removal of human immunodeficiency virus by an 0.04 μm membrane filter. *J. Acquir. Immune. Defic Syndr. Hum. Retrovirol.* 8:64–65.

Oshima, K. H., T. T. Evans-Strickfaden, A. K. Highsmith, and E. A. Ades. 1995b. The removal of phage T1 and PP7, and poliovirus from fluids with hollow-fiber ultrafilters with molecular weight cut-offs of 50,000, 13,000 and 6,000. *Can. J. Microbiol.* 41: 316–322.

Oshima, K. H., T. T. Evans-Strickfaden, A. K. Highsmith, and E. A. Ades. 1996. The use of a micropourous polyvinylidene fluoride (PVDF) membrane filter to separate contaminating viral particles from biologically important proteins. *Biologicals* 24:137–145.

Oshima, K. H., A. K. Highsmith, and E. A. Ades. 1994. Removal of influenza A. virus, phage T1 and PP7 from fluids with a nylon 0.04 μm membrane filter. *Environ. Topical Water Qual.* 9:165–170.

Pall Corp. 1995. Validation guide for Pall Ultipor® VF™ grade DV50 Ultipleat™ AB style virus removal filter cartridges. Publication TR-UDV50. East Hills, N.Y., USA: Pall Ultrafine Filtration Company, pp. 1–22.

Pall Corp. 1998. *Validation guide for Pall Filtron® Omega™ 300 K VR (virus reduction) screen channel membrane cassettes.* Northborough, Mass., USA: Pall BioPharmaceuticals, pp. 1–38.

Para, M. 1965. An outbreak of post-vaccinal rabies (rage de laboratorie) in Fortaleza, Brazil, in 1960. Residual fixed virus as the etiological agent. *Bull. World Health Organ.* 33:177–182.

Pasi, K. J., and F. G. H. Hill. 1989. Safety trial of heated factor VIII concentrate (8Y). *Arch. Dis. Child.* 64:1463–1467.

Peerlinck, K., J. Arnout, J. G. Gilles, J. M. Saint-Remy, and J. Vermylen. 1993. A higher than expected incidence of factor VIII inhibitors in multitransfused haemophilia A patients treated with an intermediate purity pasteurized factor VIII concentrate. *Thromb. Haemost.* 69:115–118.

Phillips, M. W. 1996. Integrity testing virus-retentive membranes. Paper presented at the PDA International Congress, 19 February, in Vienna.

Prince, A. M., W. Stephen, and B. Brotman. 1983. β-propiolactone/ultraviolet irradiation: A review of its effectiveness for inactivation of viruses in blood derivatives. *Rev. Infect. Dis.* 5:92–107.

Prince, A. M., B. Horowitz, M. S. Horowitz, and E. Zang. 1987. The development of virus-free labile blood derivatives–a review. *Eur. J. Epidemiol.* 3:103–118.

Prowse, C. V., and A. E. Williams. 1980. A comparison of the in vitro and in vivo thrombogenic activity of factor IX concentrates in stasis (Wessler) and non-stasis rabbit models. *Thromb. Haemost.* 44:81–86.

Prudouz, K. N., and J. C. Fratantoni. 1994. Viral inactivation of blood products. In *Scientific basis of transfusion medicine,* edited by K. C. Anderson and P. M. Ness. Philadelphia: Saunders, pp. 852–871.

Prusiner, S. B. 1984. Prions–novel infectious pathogens. *Adv. Virus Res.* 29:1–56.

Rabenau, H., V. Ohlinger, J. Anderson, B. Selb, J. Cinatl, W. Wolf, J. Frost, P. Mellor, and H. W. Doerr. 1993. Contamination of genetically engineered CHO cells by epizootic haemorrhagic disease virus (EHDV). *Biologicals* 21:207–214.

Roberts, P. 1997. Efficient removal of viruses by a novel polyvinylidene fluoride membrane filter. *J. Virol. Methods* 65:27–31.

Santos, J. A. L., M. Mateus, and J. M. S. Cabral. 1991. Pressure driven membrane processes. In *Chromatographic and membrane processes in biotechnology,* edited by C. A. Costa and J. S. Cabral. Dordrecht, The Netherlands: Kluwer Academic Publishers, pp. 177–205.

Sato, T., T. Noda, S. Manabe, T. Tuboi, S. Fujita, and N. Yamamoto. 1994. Integrity test of virus removal membranes through gold particle method and liquid forward flow test method. In *Animal cell technology: Basic and applied aspects,* edited by T. Kobayashi, Y. Kitagawa, and K. Okumura. Dordrecht: The Netherlands: Kluwer Academic Publishers, pp. 517–522.

Schreiber, G. B., M. P. Busch, S. H. Kleinman, and J. J. Korelitz. 1996. The risk of transfusion-transmitted viral infections. *N. Engl. J. Med.* 334:1685–1690.

Sekiguchi, S., K. Ito, M. Kobayashi, H. Ikeda, T. Tsurumi, G. Ishikawa, S. Manabe, M. Satani, and T. Yamashiki. 1989. Possibility of hepatitis B (HBV) removal from human plasma using regenerated cellulose hollow fiber (BMM). *Membrane* 14:253–261.

Shadle, P. J., P. R. McAllister, T. M. Smith, and A. S. Lubiniecki. 1995. Viral validation strategy for recombinant products derived from established animal cell lines. *Cytotechnol.* 18:21–25.

Shah, K., and N. Nathanson. 1976. Human exposure to SV40: Review and comment. *Am. J. Epidemiol.* 103:1–12.

Simmonds, P. 1998. Transfusion virology: Progress and challenges. *Blood Reviews* 12:171–177.

Wang, Y. J., S. D. Lee, S. J. Hwang, C. Y. Chan, M. P. Chow, S. T. Lai, and K. J. Lo. 1994. Incidence of post-transfusion hepatitis before and after screening for hepatitis C virus antibody. *Vox Sang.* 67:187–190.

Waters, T. D., P. S. Anderson, Jr., G. W. Beebe, and R. W. Miller. 1972. Yellow fever vaccination, avian leukosis virus, and cancer risk in man. *Science* 177:76–77.

Williams, M. D., B. J. Cohen, A. C. Beddall, K. J. Pasi, P. P. Mortimer, and F. G. Hill. 1990. Transmission of human parvovirus B19 by coagulation factor concentrates. *Vox Sang.* 58:177–181.

Wilson, G. S. 1967. *The hazards of immunization.* London: University of London, The Athlone Press.

7

CLEANING AND VALIDATION OF CLEANING IN BIOPHARMACEUTICAL PROCESSING: A SURVEY

Jon R. Voss

KMI Systems

Robert W. O'Brien

Biopure Corporation

Cleaning and cleaning validation remain major concerns both for product quality and from a regulatory perspective. Cleaning has always been an important factor in the drug manufacturing life cycle, but it has only been during the past 5 to 7 years that the emphasis on cleaning has reached such a state of heightened anxiety. For the biotechnology industry, the flurry of cleaning "awareness" began in the summer of 1992, when the first commercial multiproduct manufacturing facilities were undergoing their licensing reviews. Also, in 1993, Justice Alfred M. Wolin of the U.S. District Court ruled on the Barr Laboratories, Inc. civil action (*U.S. v. Barr Laboratories, Inc.* 1993), regarding their failure to validate several processes adequately, including their cleaning processes.

This chapter discusses the concepts involved in cleaning chemistry, cleaning systems, and the validation of cleaning processes. Due to the tremendous variation in cleaning agents, cleaning equipment, and cleaning processes, the information contained herein is intended to serve as a guide for those involved in developing cleaning programs and in validating cleaning of equipment and cleaning processes.

CLEANING

Cleaning has both a direct and an indirect effect on product quality. The obvious direct effect is contamination. If soils are not removed from product contact surfaces, they can potentially become sources of direct product contamination. Indirect contamination also can be a result of inadequate cleaning. Biotechnology soils serve as ideal growth media for many microorganisms. Since most biologically derived products utilize aseptic processing, with filtration as the primary means of sterility assurance, microbial control/minimization is an important aspect of cleaning and cleaning validation. Failure to clean surfaces can leave residues that encourage microbial growth and lead to indirect product contamination.

Uniqueness and Types of Soils

Soils associated with biotechnologically derived products are substantially different from those derived from traditional pharmaceutical products. By their very nature, the soils derived from biotechnology processes tend to consist of complex proteins and protein fragments. Proteinaceous soils typically adhere tightly to the surfaces of equipment, particularly when dry, and tend to serve as excellent sources for microbial growth. Combined with typically poor solubility, when compared with traditional pharmaceutical products, biotechnology cleaning and cleaning chemistry is much different than that found in the traditional drug industry.

Chemistry and Agents

The chemistry of cleaning in the biotechnology industry is similar to that of the dairy industry. Most cleaning agents used today, indeed most cleaning approaches used in the biotechnology industry, were derived from techniques used in the dairy industry. This seemingly unusual tie to what might appear at first to be an archaic industry, when compared to the new "high tech" biotechnology industry, is actually very logical.

The dairy industry relies heavily on sanitary processing equipment, and while not aseptic, their processing requirements are often more stringent than that found in many biotechnology plants. Like the biotechnology industry, the dairy industry is highly automated and uses state-of-the-art processing equipment. As found in most competitive, high-volume, low-profit margin industries, technology is used to reduce costs and increase profit. Indeed, all of the current top five clean-in-place (CIP) equipment vendors began by supplying the dairy industry with cleaning systems.

The soils of the biotechnology and dairy industries are comparatively similar too. Dairy products contain a complex mixture of proteins, free fatty acids, and polysaccharides not unlike the precursors to most biotechnology fermentation and cell culture processes. The similarity of the two industries and, consequently, their processes is perhaps most clearly seen when one recalls that cows, goats, and pigs are presently being routinely used to produce and express biotechnologically derived proteins in their milk!

The cleaning chemistry for biologically derived products has necessarily been designed to accommodate the requirements of the soils; since the soils are unique, so are the cleaning agents. Cleaning agents for biologically derived products tend to be designed to break up and solubilize the proteins, protein fragments, sugars, and fatty acids found in biotechnology process residues.

Cleaning agents tend to fall into a few select categories based on function (Table 7.1). The most commonly used agents in the biotechnology industry tend to be alkaline based. Alkaline cleaning agents are effective protein inactivators, fatty acid saponifiers, and general dissolving agents, which makes them ideal general purpose cleansers. Alkaline cleaning solutions are frequently chlorinated to

Table 7.1. Typical Cleaning Agents

Cleaning Agents	Concentration
Acidic	
Acetic acid	100–200 ppm
Peracetic acid	100–200 ppm
Phosphoric acid	1,000–2,500 ppm
Basic	
Sodium hydroxide	1,500–7,500 ppm
Sodium hypochlorite	25–50 ppm
Solubilizing detergents	Based on manufacturer's recommendations

improve protein denaturation and their microbial inactivation capabilities.

Acidic agents are effective at breaking down proteins (peptizing) and dissolving complex sugars. Organic acids used as cleaning agents include citric, tartaric, and acetic acids. The most common inorganic acids used for cleaning are phosphoric and nitric acid. Blends of these acids are frequently used to "chase" alkaline cleaning agents to effect neutralization. Chasing an alkaline cleaning cycle with an acidic rinse tends to reduce the volume of water necessary to remove the alkaline residues and, therefore, has an economic as well as process benefit.

The choice of a cleaning agent—its concentration, contact time, temperature, and delivery mechanism—should be based on the chemistry of the soils, the nature of the process, and the type of processing equipment. Development studies need to be conducted prior to any validation work being conducted. These studies should be well documented, as they form the basis for the cleaning validation program itself. Without them, the ability to select acceptance criteria is restricted dramatically and can result in delays and process failures.

Development studies should indicate why an agent was chosen, what the test plan is, and clearly document the results of the testing. These data then support the rationale for future cleaning agent use and selection.

Unfortunately, cleaning agents are often chosen based purely on vendor recommendations or from a poorly documented prescale-up process. Failure to choose a cleaning agent properly can easily result in start-up problems, validation difficulties, and process quality issues.

Cleaning Approaches for Equipment

The cleaning approaches for equipment used in the biopharmaceutical industry have been derived from methods developed in the traditional pharmaceutical industry. The most common of these approaches, CIP, is favored because it lends itself well to automation. A basic CIP system is typically of a packaged skid design, which contains a chemical dosing pump, a solution makeup tank, temperature and pH/conductivity control, a recirculation pump and a programmable logic controller (PLC). This equipment, as the name implies, allows for soiled processing equipment to be cleaned where it is used; it requires no disassembly and transport to a cleaning station. CIP systems also can be very sophisticated, containing multiple tanks, pumps, and control loops.

Similarly, clean-out-of-place (COP) systems are skid design packages used for cleaning small, intricate, portable, and/or removable processing parts not easily cleaned-in-place. COP also can be automated and is often used in place of CIP to reduce the cost associated with the implementation of a cleaning system that could involve multiple tanks, several pumps (delivery and recirculation), lots of piping, and require interfacing between multiple controllers.

Manual cleaning is often considered the "ugly duckling" of cleaning processes. Once considered difficult to control and impossible to validate, manual cleaning is beginning to be recognized for what it is—a necessary, useful, and validatable cleaning process.

Because manual cleaning requires human interaction, variability in the overall effectiveness limits its reproducibility. This human factor means that operator training becomes the key to success and reproducibility. Operators must carefully follow well-documented procedures, and the cleaning process must be robust enough to allow for variances in operator interaction.

Clean-in-Place

CIP is the preferred method of cleaning for both traditional pharmaceutical and biopharmaceutical industries. CIP is preferred because it is relatively easily automated, and automation implies that control is repeatable. Repeatable control is the panacea of every validation and quality-minded person.

CIP systems may be either fixed or portable. Fixed CIP systems are typically more complex than portable systems and require careful planning in order to design a system that can be used to clean multiple, fixed, and often large pieces of equipment. Such systems usually have relatively sophisticated control systems to dose chemicals and to make up, deliver, and route solutions to the equipment or system to be cleaned.

Portable systems are often utilized when the design of the equipment makes it difficult or impractical to supply cleaning solutions or when the equipment and/or piping systems were not designed to accommodate CIP operations. Portable systems may be either custom or "package" systems (vendor's basic off-the-shelf unit). While fixed CIP systems may utilize one or more programs servicing many different cleaning "circuits," portable systems typically only have one or a limited number of cleaning cycles. Portable systems usually only are designed to make up solutions and deliver the solutions to their intended target equipment. They do not typically control the delivery, solution routing, or return valving used to sequence the cleaning process.

Clean-out-of-Place

When CIP is not possible or practical, COP is often the next best thing. COP is often used to clean small parts and small items of equipment that are either too small, complex, or inaccessible to be cleaned via CIP. Depending on the system, a COP system may be more or less automated.

COP systems can vary tremendously in their overall sophistication and application. They can be represented by a bucket with a brush or, alternatively, as a highly automated multitank cleaning system.

Sophisticated COP systems use automated control to measure incoming source water, mix cleaning chemicals, and temper and deliver cleaning solutions to one or more portable vessels or tanks. These tanks, called parts washers, are preloaded with the various equipment parts to be cleaned with the cleaning solutions being recirculated through the tank. Multistage processing with different chemicals is possible, and efficient cleaning operations can be effected.

Manual Cleaning

Used extensively for difficult-to-clean small components, those that are difficult to clean-in (or out)-of-place, manual cleaning is a necessary part of the overall cleaning picture. Manual cleaning is often used when special cleaning action is necessary. Small components, fragile components, or components that were not cleaned effectively using the turbulent solution flow methods of CIP or COP often must be manually cleaned utilizing a "scrubbing" action or prolonged soaking to effect cleaning.

Cleaning Process Design

Equipment Design (for cleanability)

In order to be able to clean a piece of equipment, it must be designed to be cleanable. Designing equipment to be cleanable is often less of a science than an art. Despite this fact, certain general design principles can be applied to improve a system's "cleanability."

Designing for cleaning requires that many factors be considered. Cleaning solution delivery, circulation, surface contact, and removal are all parameters that must be analyzed and reviewed relative to their impact on the ability to clean soils from process equipment. Delivery of cleaning solutions to the soiled equipment may come from simple "passive" contact, such as through tank or vessel "flooding," to more "active" forms, such as the use of spray balls or wands. Somehow, cleaning solutions must be delivered to all product contact surfaces. Typically, the more active the delivery, the more effective the cleaning, although passive techniques are often more than adequate to achieve the necessary level of cleaning for certain applications.

Most equipment used in the biopharmaceutical industry is composed of relatively strong, inert materials. Stainless steel, glass, Teflon®, and a handful of other polymeric compounds are the construction materials of choice. These materials have generally been accepted by the industry and regulatory bodies because they are inert, nonreactive, and for the most part easily cleanable.

Surface finish is directly related to surface "cleanability." Nonporous, smooth finishes are more cleanable than porous, rough finishes. Technically, the reason for the difference in cleanability is tied to surface finish and area. A rough surface will yield more surface area than a smooth surface for the same amount of soil to bond to. Surfaces should be as smooth as practically possible. This does not mean that all product contact, stainless steel surfaces, for example, must be electropolished to be cleanable. It does mean, however, that the surface finish must be considered when equipment is being designed, as well as compatibility of materials, product soils, cleaning chemicals, and the cleaning process. A roughness average (RA) of 10 to 20 μin. (microinches) is typically considered appropriate for electropolished, wetted process surfaces. Appropriate choices must be made to optimize all of these parameters.

In order to be cleanable, equipment must be capable of draining freely. Proper drainage is a key element of cleaning design. Improper drainage can result in nonprecipitated or dissolved soils adhering to the surfaces that are to be cleaned. These soils, if left to dry, are often harder to clean than the originally soiled equipment.

Piping systems should be sloped to produce a freely draining pipe. Typically, a pipe slope of 1.0 in. rise per 10 ft run is considered minimally acceptable for most water-based solutions. In addition, piping size should be considered so as to not induce a "capillary effect."

Circuit Design

Cleaning circuits or paths must be carefully designed to provide for delivery of the cleaning solutions, circulation or contact of the solutions with the surfaces to be cleaned, and drainage and return of cleaning solutions. The design process must take into account physical, logical, and practical factors. The proper blend of these factors is the "art" of circuit design.

Physical factors that affect circuit design include the sizing of equipment (e.g., piping, pumps, vessels) and the distance or location from the cleaning system. Pumps must be sized in order to deliver the appropriate flow rates necessary to clean the largest and smallest equipment item in a circuit or series of circuits. The common rule of thumb applied to achieve adequate cleaning based on the principles of fluid mechanics and turbulent versus laminar flow is a flow velocity of at least 5 ft/sec. While somewhat arbitrarily determined and shown to be only a general rule, the 5 ft/sec rule is useful as an initial design base measure for sizing piping and pumps.

Pipe diameters and lengths must similarly be sized to accommodate the maximum and minimum flow rates delivered by the cleaning system. This can be difficult to achieve with a single cleaning system, as the ability to choose a delivery pump that can vary its delivery pressure to accommodate the desired flows is limited.

Cycle Development

A cleaning cycle can be as simple as a manual soak and rinse or as sophisticated as a highly automated, multifluid, multipath design. When discussing CIP/COP, the latter is typical. In either case, the proper design of a cleaning cycle should take into account certain factors and parameters. First, the cycle design should logically consider the complexity of the equipment or "circuit" to be cleaned in relation to its potential for soil holdup or retention. What size is it? Does it contain numerous equipment items of varying shapes? Are there any hard to reach solution contact areas identified during the circuit design phase?

The type of anticipated soiling conditions is the second factor that requires consideration. Along with this is knowledge and understanding of the chemical type(s), solution concentrations, and temperatures required to break down and remove the soil. This knowledge is typically gained during cleaning development studies. If properly conducted, these studies become the basis for good cycle design.

Once the proper chemical parameters are known, it becomes a relatively easy process to determine the remaining three factors of cycle design: the number and types of fluids, the fluid path (flush/rinse or recirculation), and the recirculation time. A typical approach used in the biotechnology industry utilizes an initial caustic solution or water flush to drain to remove the heavy bulk of soil. This is followed by recirculation of multiple solutions used for soil breakdown and/or chemical neutralization. Multiple water rinse cycles then follow. In contrast to the well-defined chemical information for soil removal gained from development studies, the fluid recirculation times and number of rinse cycles necessary must be arbitrarily picked due to the unknowns of soil buildup and retention levels in any given "circuit." These cycle parameters can only be determined with proper testing of the cycle design.

Cycle Development Testing

Cycle development testing sets the groundwork for cleaning validation. Qualification of effective performance of a cleaning cycle is typically performed in three phases. Phase 1, cycle development, is used to establish and set CIP circuit recipes. During this phase, rinse samples are analyzed to verify the proper number of rinse cycles and rinse cycle durations required.

Phase 2, final cycle development, is designed to verify that the previously established cleaning cycle is effective. Phase 2 testing may utilize any appropriate combination of rinse water analysis, visual inspection, and surface analysis (swabbing) to verify this effectiveness.

Phase 3, circuit performance qualification, is intended to verify that the previously tested cleaning cycle is consistently effective. Phase 3 testing involves soiling and holding the equipment to be cleaned under perceived "worst-case conditions," such as high soil concentrations and the longest hold time anticipated prior to the start of cleaning.

Special Cleaning Applications

Typical biotechnology equipment used in the industry includes fermentors, bioreactors, buffer tanks, media tanks, filters, chromatography columns, filling systems, and their associated support equipment, including pumps, valves, and piping. These categories are discussed below.

Piping, Tanks, Pumps and Valves

The cleaning of piping, tanks, pumps, and valves is relatively straightforward—the emphasis on cleanability must be taken into account at the time of design. For piping, this means that it is sized adequately to receive and maintain minimum turbulent flow velocities of 5 ft/sec, it is sloped properly for drainage, dead legs are minimized, and the surface finish is of adequate "smoothness."

Cleaning design considerations for tanks should include adequate fluid velocity and total surface coverage from spray ball nozzles, such that tank lids are completely contacted, dip tubes do not adversely block the spray coverage, and all inlet and outlet tank nozzles receive complete contact within their necks. Consideration should also be given to the tank's vent filter to ensure that it is of adequate size and material so as not to "blind" from saturation during cleaning operations. Vent filters constructed of hydrophobic membranes are typically used for this application.

Two different types of pumps, centrifugal and positive displacement, are used for sanitary applications in the biotechnology industry. "Sanitary" pumps, in theory, do not allow impurities, such as bacteria, to enter the system and contaminate the product.

Examples of positive displacement pumps are rotary lobe, metering peristaltic, metering diaphragm, diaphragm, and reciprocating piston type (Cesar 1994). When choosing a pump type for a specific process application, the following details are of extreme importance:

- Product viscosity.

- Product temperature.

- Shear sensitivity of the product.

- Materials of construction for the pump.

- Surface finish for the pump.

- CIP cleanability of the pump.

CIP cleanability is typically more difficult to achieve, though not impossible, for positive displacement pumps than for centrifugal pumps. Positive displacement pumps tend to have internal dead pockets around lobe splines and seal areas that trap product, whereas centrifugal pumps can be manufactured with special, double seal shaft designs and flush systems. Positive displacement pumps may also cause problems with lobe interference of the CIP solution recirculation. If a process application requires a positive displacement pump, ensure that it has CIP capability, otherwise it will require disassembly and cleaning-out-of-place.

The design considerations used when choosing a process valve in the biotech industry are no different than those detailed above for selecting a process pump. When choosing a valve type for a specific process application, the product's attributes and processing parameters, the valve materials of construction and surface finish, as well as consideration of CIP cleanability are all important.

Filters

Filtration is a common process operation in the biopharmaceutical industry; it is primarily used to accomplish some aspect of product purification. Its use may range from the straightforward removal of foreign particulate matter by microporous filtration to the complex separation of selected proteins from a mixture of proteins by ultrafiltration on a molecular weight basis. The types of filters used may be disposable polymeric membranes or reusable depth filters, such as diatomaceous earth candles. The latter are commonly used for early process separations to conserve on cost relative to disposable membrane filters. Final product sterilizing filtrations are performed using disposable 0.2 μm membrane filters.

Relative to cleaning, the disposable membrane filters are discarded after use and, therefore, require no cleaning. If the filters are integral to a disposable housing, the entire unit is discarded. If the filters are installed in a stainless steel, reusable housing, the housing must be disconnected, the filters removed, and the housing connected to a CIP system. Alternatively, the housing may be disassembled for off-line cleaning either by hand or COP.

Reusable filter cleaning takes on many forms depending on the type of filter and its intended use. Flushing or rinsing is by far the most common means of cleaning and is frequently used as the only means of cleaning. Flushing may be performed either in the forward or reverse flow direction, again depending on the type of filter. Reverse flow (backflushing) versus forward flow is generally preferred if compatible with the filter. The choice of cleaning agent(s), concentration(s), and temperature(s) are all dependent on their compatibility with the filter membrane material and support structure.

Rinsing is often followed with a chemical soak phase to inactivate bound proteins and/or reduce bioburden. Sodium hydroxide is one of the most common soaking agents used, although chlorinated solutions are sometimes used, depending on the filter's tolerance. Testing should be conducted to determine final rinse water quality, after soaking, for pH, Cl^-, endotoxin, bioburden, and total organic carbon (TOC) levels as deemed appropriate.

Besides assessing final rinse water quality, determining whether a filter has been adequately cleaned can best be determined by measuring the flux rate and comparing the values obtained to the values supplied by the filter manufacturer. Typically, manufacturer-stated flux rates are determined with water. It is important to also measure and assess filter flux rates, after cleaning, with the filtered product to ensure proper process performance.

In addition, the impact, if any, the cleaning cycle has on filter integrity should be assessed. If a filter is to be reused, studies should be performed to determine the number of times a filter can be cleaned and reused and the conditions that dictate replacement of a filter.

Chromatography Columns

If filtration is "rough" purification, then chromatography is "fine" purification. Used heavily in the biopharmaceutical industry to purify the desired protein by removing contaminants or process by-products, chromatography systems present several unique cleaning challenges.

Chromatography columns are often reused many times before they are "changed out" or repacked. This is partially due to the expense of the column matrices, but more appropriately because the nature of chromatography is a regenerative process. Once a column has been used, it is typically cleaned, regenerated, and prepared for reuse, often within a few hours. If not reused immediately, columns are frequently stored "wet" in a sanitizing solution and then cleaned prior to reuse.

Cleaning a chromatography column is similar to that of a filter. Flushing with a chemical solution is the most common method employed. The type and concentration of cleaning solution(s) used vary depending on the medium matrix and purity level of the chromatography feed stream. If the feed is relatively dirty—e.g., having a high bioburden load—the solution utilized for the cleaning step should be a strong chemical solution. A relatively clean feed stream may require utilizing only a high salt buffer solution for each cycle cleaning step, followed by a more stringent cleaning (concentrated acid wash for example) after a specified number of cycles. In any case, the cleaning is designed to remove bound and free contaminants from the column matrix and other product contact surfaces (i.e., the column itself).

In addition to column washing, a chemical sanitizing recirculation cycle is often utilized to effect viral inactivation of any bound column contaminants (White et al. 1991). Testing must be conducted to demonstrate not only the effectiveness of the inactivation cycle but also assess the impact the cleaning and sanitization may have on the column's separation process capability and useful life.

Facilities

Cleaning and sanitization of the exterior surfaces of all process equipment, floors, walls, ceilings, and drains in all controlled environmental processing areas should be conducted on a periodic basis so as to control and minimize the potential for product contamination. The frequency of this cleaning depends on the criticality of the processing being conducted and the equipment requirements for this processing. A typical, regimented, facility cleaning and sanitization program would require daily treatments for drains, twice weekly for floors, twice monthly for walls, and quarterly treatments for ceilings and equipment surfaces. In contrast to this type of schedule would be the cleaning and sanitization of a critical processing area, such as a filling suite. A critical area warrants a thorough cleaning/sanitizing both before and after a fill operation, in addition to the routine periodic cycle.

Chemical solutions used for this type of cleaning and sanitization are typically a 250 ppm sodium hypochlorite solution for drains and two types of disinfectants alternately used for all other surface applications. The industry-recommended practice for disinfectants is to use a quaternary solution in an alternating fashion with a phenolic agent to counterattack any potential for a microbial resistance developed from the repeated use of one disinfectant.

Passivation

It is necessary to briefly discuss passivation in this chapter on cleaning validation because the two share a common link, surface finish of stainless steel. The use of stainless steel as a product contact material is commonplace in the biotechnology industry; it is constantly being exposed to long-term corrosive media, such as Water for Injection (WFI), clean steam, and various chemical solutions. The result of this constant exposure may be corrosion of the stainless steel surface, as well as pitting and the development of cracks and crevices in the welds. Obviously, the more imperfect the surface, the more area for product entrapment, and the more difficult to clean effectively.

To combat this ever-occurring corrosion process, a chemical solution treatment—passivation—is periodically employed to form a "passive" surface film layer that acts as a barrier to further corrosion. The state in which stainless steel exhibits a very low corrosion rate is known as passivity (Coleman and Evans 1990). The passivation process also removes oxides, free iron, and other contaminants. Typically, passivation of stainless steel piping is conducted prior to operational use, as the act of welding deteriorates the passive film established, even from electropolishing. Passivation also may be required from time to time as oxides develop on internal pipe surfaces during use.

CLEANING VALIDATION

What Is Cleaning Validation?

Cleaning validation is the process of demonstrating and documenting that a cleaning process is effective and reproducible. Besides being a regulatory requirement, validation is important because it challenges the cleaning process and can help to identify potential problems that might not be identified with routine in-process testing.

The Basics

Planning

As with any type of project, proper planning is essential. Planning is more important for cleaning validation than any other form of validation. This is because cleaning validation requires the interaction of many different groups in a tightly controlled time sequence.

The development of a cleaning validation plan is one of the most valuable activities for ensuring the success of a cleaning validation effort. A good plan should, as a minimum, contain the following information:

- Description of what the plan contains.
- Description of the equipment being cleaned.
- Description of the cleaning strategy to be employed.
- Description of the validation approach/philosophy.
- Description of acceptance criteria.

Protocols

Traditionally, validation protocols come in at least three categories: installation qualification (IQ), operational qualification (OQ), and process or performance qualification (PQ). A protocol is simply a document that describes how a test or series of tests should be performed and how the testing results will be recorded. Additionally, the protocol must provide the pass/fail criteria used to assess the test results. A protocol need not be long, but it should be clear and well thought out. A poorly developed protocol can easily waste a tremendous amount of time and/or money if the test fails due to poor design or planning.

While much has already been written about IQ and OQ, it should be noted that for an automated cleaning system, IQ and OQ testing is an important part of the cleaning validation program. IQ and OQ provide the foundation on which PQ rests; if that foundation is weak, PQ will be correspondingly weak. For an automated system, IQ should verify that the system has been installed properly, that the software is structurally sound, and that the configuration has been "locked down" and documented. OQ should verify that the automated system functions as designed and specified. Operational sequence testing and timer operation should be evaluated, as well as error detection and alarm actions. Sequence testing verifies that the programmed cleaning cycle executes in the order prescribed. During sequence testing, tables indicating valve positions for proper operation are typically compared to actual field positions. Similarly, OQ execution timer lengths may be verified as well as alarms verified for proper operation.

The structure built on the IQ/OQ foundation is PQ. PQ challenges the cleaning process. In order for PQ to be useful, it should provide a reasonable level of testing data to substantiate the claim that the cleaning process is effective and that the process is reproducible. The PQ protocol should be clear and concise. It should describe or reference the cleaning process and the equipment being tested. The testing approach (or design) should be presented and, most importantly, the rationale for the testing should be explained. A good cleaning protocol will clearly explain how and when samples will be obtained and provide a matrix or schedule that identifies where and how in the process the samples will be taken.

There are currently at least two different philosophies (based on organizational management) for the design of a cleaning process qualification study. One approach is to develop a single protocol that covers all cleaning circuits; the other approach is to develop a separate protocol for each circuit. Each approach has its advantages; the single protocol approach reduces the review/approval time, but the protocol itself is difficult to comprehend due to its complexity. The multiprotocol approach creates more protocols to review/approve, but it is easier to understand what each one is accomplishing.

The PQ protocol is important for another less obvious reason. The protocol is a guide that explains the roles and responsibilities of the other groups that will be involved in or support the cleaning validation studies. Cleaning validation is probably the most lab-intensive validation work that exists—even more demanding than traditional process validation. The PQ protocol typically defines the sample plan that lab groups will use to determine what additional requirements for support may arise as a result of the plan. Similarly, the process of running multiple cleaning cycles often requires a significant amount of manufacturing support to perform the cleaning process itself. The protocol helps define these tasks so that the supporting groups can plan accordingly and prevents the inevitable confusion if such coordination is not carefully prearranged.

Study Execution

Even if the testing protocols are well written, cleaning validation study execution is a resource-consuming task that must be carefully planned. This is because of the number of interdependent tasks that require the interaction of many different organizational groups:

- Sample collection set up (sample kits)—validation.
- Sample labeling and submission forms—validation.
- Laboratory preparation/setup—test lab.
- Manufacturing record setup—manufacturing.
- Manufacturing equipment setup—manufacturing.
- Manufacturing process operation—manufacturing.
- Sampling execution—manufacturing/validation.
- Sample submission—manufacturing/validation.
- Sample testing—test lab.
- Sample results data collection and analysis—test lab/validation/quality.

With careful planning, however, the execution phase can be achieved with a reasonable level of control. By including a complete sample matrix or table in the PQ protocol, the labs and manufacturing groups will know exactly what kind of sample load they can expect to receive; when combined with an execution schedule, they can plan their work load and resource requirements accordingly. Providing preprinted sample labels and sample submission forms will reduce the chances of samples being missed during the execution phase.

Reports

Validation reports are one of the most "visible" documents of the entire validation effort and, hence, one of the most important. If the protocol was well written, the writing of the report should be a very straightforward process.

A validation report is not unlike a scientific paper. A good report will include the following components:

- Abstract/executive summary.
- Introduction.
- Methods.
- Results.
- Discussion.
- Conclusion.

Much of the information necessary to write the report should be included in the protocol. For instance, the methods used should be identical to those described in the protocol (any deviations must be reported, and their impact assessed). Similarly, the results section should be just a repeat of the sample matrix with a column added for the actual test results.

The bulk of the report should focus on a careful analysis of the results—the discussion section. This section of the report is important because it analyzes the results in terms of the purpose of the test and the acceptance criteria. The analysis should clearly answer the following two questions: "Why is this result important?" "What does it prove?"

The discussion section should also clearly address the impact of any test results that failed to meet the protocol acceptance criteria. Should a result fail, its impact must be addressed. A failed test result does not necessarily mean that the validation failed and that the cleaning process is ineffective or nonreproducible. Test results often fail due to operator error, sampling problems, or poor study design. While troubling, testing failures do occur and must be expected.

Cleaning Equipment Validation

Installation Qualification

IQ testing is all too often a frustrating exercise that frequently produces pounds and pounds of documents of dubious worth. The reasons for this are both complex and simple. The simple explanation is that IQ testing is a relatively "brainless" activity that requires little deep thought or analysis; it is primarily a bureaucratic activity. The complex explanation may be related to the fact that validation is a poorly defined activity and open to varied interpretations (Lord 1995). Often, various interests (many by poorly informed consultants) are served by producing incredibly complex documents.

What then should an IQ do? A successfully completed IQ should provide a reasonable level of assurance that the system being qualified has been installed properly. Proper installation is typically verified by comparing the installed system with the applicable specifications. Installation specifications for utility connections as determined by the manufacturer, vendor, system integrator, and so on should be compared with the "as-built" or "as-installed" system documentation. Often, independent verification is not necessary; acceptance based on qualified vendor documents may be sufficient. Exceptions to this generalization typically apply when a utility is critical to the operation of the system or could impact product quality.

One of the most important benefits gained through the IQ process is the review and collection of supporting documentation for the system. Standard Operating Procedures (SOPs), drawings, manuals, instrumentation lists, and so on are identified, located, and compared to an expected list of necessary documentation. If not performed as part of IQ, this documentation is often lost or never received, thus making start-up and continued operation difficult.

Often, the IQ process serves the useful purpose of collecting in one location all of the documentation for a system. The benefits of this effort may be lost, however, when the validation files are kept locked up as "controlled documents." This prevents those who need the documents most (mechanics and users for instance) from having ready access to the very information they need. Similarly frustrating is the fact that these document collections often become obsolete as soon as they are generated because the update process rarely includes these documents.

IQ work should occur prior to OQ work, as the purpose of IQ is to provide assurance that the system has been installed and is in an acceptable state to start up. Once started up, the system can then be qualified via OQ.

It should be thought out in advance what parts of IQ must be complete prior to the commencement of OQ activity. If not planned for and agreement reached among the responsible groups, delays in start-up and validation tasks can occur. It may be considered acceptable for OQ to begin as soon as IQ execution has been completed but before the IQ report has been reviewed and approved. The decision for what is necessary is company specific but should be based on sound logic, with the consequences of the decision being well understood. These aspects of planning are important to the overall validation of the facility or process start-up; if thought out in advance, it will save time and money.

Operational Qualification

OQ testing is designed to verify that the system or equipment item will reliably operate as designed and/or specified. The determination of what is considered "proper operation" is, of course, the crux of the issue. Minimally, verification of proper operation should include tests of the key system functions–those functions that the system was designed to control or monitor. The degree to which functional testing is applied to a given system function depends on the importance of the function itself. Certain functions will require relatively minimal testing–does the function work under expected conditions? Usually however, more extensive testing is required to verify that the function will work under adverse or unexpected conditions.

OQ testing for cleaning systems typically involves at least five different types of testing: sequence testing, alarm testing, interlock testing, operator interface testing, and control loop testing.

Sequence testing verifies the operational sequence of the CIP/COP control system. CIP/COP control systems typically employ a PLC to effect the control of the system pumps, valves, and sensors. The PLC, in turn, is programmed to perform specific output actions given a set of known input conditions. The program(s) that perform these actions typically are based on a set of matrices or sequence tables that specify the valve positions (open/closed) and pump actions (on/off) for a specific set of input conditions. OQ sequence testing typically involves the verification of the sequence table. Verification is usually accomplished by forcing the input conditions for each control sequence and verifying the output functions against the sequence table.

Alarm testing, as the name implies, verifies the operation of alarms associated with the system. Alarms usually fall into one of two categories–alarms designed to protect the equipment (high pressure, etc.) and alarms intended to notify the operator of a failure condition (supply tank low level, etc.)–that serve to protect the product/process or operator. When determining alarms to be tested, consideration should be given to the impact of the alarms, and priority should be determined as follows:

- Process or safety specific (highest priority): Alarms in place to protect product and/or personnel safety. Failure will compromise product or endanger personnel.

- Equipment protection (middle priority): Alarms in place to protect equipment from damage.

All process or safety specific alarms and selected equipment protection alarms are tested.

Interlocks may be mechanical or part of the electrical/mechanical control system. Interlocks typically either serve as preventive measures (turn the pump off before the tank is empty and the pump cavitates) or to initiate an activity (when the start button is pressed, the tank is full, and the CIP circuit selector switch is in the "on" position, then start the cycle). Interlock testing may be done as part of sequence testing or as a separate activity. Like sequence testing, interlock testing involves the forcing of inputs either through simulation or through "expected" operating conditions and verifying that the interlock action is effected.

Although *operator interface* testing is receiving less emphasis with the advent of more pragmatic approaches to computer validation, it is still an important part of the OQ process. Interface testing should verify that the basic interface functions as expected—menus are displayed as expected and indicators and keys operate properly. If the interface expects operator input (e.g., the entry of setpoints), then the interface should be tested to verify that an unexpected input (e.g., out-of-range conditions) does not result in problems that would affect system operation or performance.

Control loop testing is one of the most important aspects of OQ testing. While sequence testing verifies the proper operation of the control system with respect to input and output (I/O) sequences, it does not verify the "analog" operation of most control loops. Typical I/O testing merely tests the discrete signals of a control loop. For example, I/O tests will verify that a signal is received to elevate temperature and that the temperature begins to rise, but it does not test and challenge the controllability of a loop throughout the process range. Temperature, liquid level, concentration, and pressure are the most common control loops found in CIP/COP systems. Critical control loops (those loops that could impact product quality if they should fail) should be challenged to verify that they control as expected and, under unexpected conditions, are capable of responding as needed. Such testing usually involves testing the loop at the setpoint to verify the ability of the system to control at the setpoint for an appropriate duration. Unexpected operational testing often is effected by forcing the control loop outside the normal operating range to determine whether the system can respond and bring the loop back into control.

Validation of the Cleaning Process: A Three-Step Approach

The validation of a cleaning process, whether automated or manual, can generally be divided into three steps. The first two steps need not necessarily be performed as part of the PQ study (they are often considered prequalification or development activities), but they often are included and provide a solid base on which to build the final PQ testing procedures.

Step 1: Cycle Design/Development

Regardless of the cleaning approach (CIP, COP, or manual) used, a cleaning procedure or cycle must be developed. Often, the cleaning cycle evolves from trial-and-error techniques or is passed along with the process as the process is moved from the lab and scaled for clinical or commercial operation. The cleaning cycle defines the times, the cleaning solutions, the cleaning sequence, and the manual procedures necessary to effect a cleaning program. A cycle most likely is unique to a particular equipment item, or it may be applicable to many equipment items in a facility. CIP systems, for example, typically have a unique cycle for each cleaning circuit. This is due in large part because each circuit is itself unique, with differing path lengths, piping diameters, and tank volumes.

When scaled up or transferred, a cleaning cycle often must be modified. The only part of a cycle that can usually be transferred is the cleaning agents; the choice of cleaning agents should have been documented and efficacy verified. The remaining parts of a cycle (the times and sequence) are so circuit or equipment specific that they must be developed at the time of initial cycle "start-up."

As much as one might expect the establishment of a cleaning sequence (valve positions, pump times, etc.) to be a rigorous, logical process utilizing automated programming techniques, the truth is that trial-and-error techniques are far more commonly needed. These procedures are used because the calculations for drain and fill times are so complex and involve so many assumptions that often the only pragmatic approach is to use a stopwatch to determine the associated times for various steps in the cleaning cycle.

Cycle development and design should include an analysis of the selection of the cleaning agents. Cleaning agent selection should be done well in advance of PQ testing; however, this is not always the case, for example, when the cleaning agent is changed based on scale-up considerations. The rationale for cleaning agent selection should be well documented, preferably in a cleaning process development report (Agallaco 1992). Selection of the agent and the parameters for its use (temperature, concentration, pH, etc.) should be documented and based on solid test data. These data are especially important for establishing acceptance criteria.

Step 2: Cycle Testing

Once a cycle has been designed, it must be tested. Initially, the testing is performed to verify that the crude cycle parameters produce acceptable results. Drain times, for instance, are tested to verify that for each circuit and drain sequence, the times allow full and free drainage.

Once a cycle's crude parameters have been tested, the ability of the cycle to remove the cleaning agents should be tested. Establishment of cycle rinse times should verify that when operated using normal operating parameters, the rinse cycle successfully removes the cleaning agent from the process equipment.

Most rinse testing involves the analysis of samples of rinse water for traces of the cleaning agent(s). Depending on the nature of the cleaning agent, pH or conductivity analysis may be used to verify rinse effectiveness.

Cycle testing provides confidence that the cleaning cycle(s) will operate in a consistent manner and can remove the cleaning agents that will be used in the cleaning process. It does not verify that the cleaning process is effective, as that requires the use of actual product or placebo.

Step 3: Process Testing (Analytical Methods, Sampling Techniques)

The final step in the validation of the cleaning process involves verification of the cleaning process under actual process conditions. This phase of testing requires the use of product or product analogues that mimic the cleaning chemistry of the actual product.

Process testing often involves several methods of cleaning efficacy verification (McArthur and Vasilevsky 1995). The most common of these are visual detection, rinse water analysis, and surface analysis (via swabbing).

Rinse water analysis is the most common test method for cleaning validation. Like visual detection, it is relatively fast and inexpensive. Unlike visual detection, rinse water analysis is quantitative. Rinse water is typically used to determine cleaning effectiveness by testing the final rinse water for product, detergent, bioburden, pyrogens, and/or other contaminants. Although rinse water is not a direct indicator of cleaning effectiveness (it has no direct ability to assess surface cleanliness), it is the most commonly used analytical method and is the primary test method for most cleaning validation programs.

In order to determine cleaning effectiveness directly, one must use either visual detection or swabbing. Visual detection is a common and useful analytical test for assessing the effectiveness of a cleaning process. Depending on the visual detectability of the process soils, visual detection is the easiest, fastest, and least expensive analytical technique that can be employed. Unfortunately, visual detection is more qualitative than other analytical methods and, therefore, is sometimes discarded as a valid analytical tool. This is a mistake because visual detection limits are better than most people would be led to believe.

To utilize visual detection effectively as a test method, the method itself must be validated. Coupons are spiked with decreasing concentrations of product or cleaning agent and allowed to dry. Several trained operators inspect the coupons under controlled lighting conditions, and the limit of detection is established. Typically, a safety factor is added to the minimum or average value obtained from the testing; then a visual detection limit is established.

One of the most recent additions to the biotechnology cleaning validation test method is swabbing. Swab analysis provides a direct measurement of cleaning effectiveness by physically removing adhered soils from the surface with a swab (Smith 1992). Like the visual and rinse water test methods, however, swabbing has its own share of problems. Swabs must be selected to be compatible with the equipment being swabbed, the solvent used to extract the soils from the surface, and the analyzing test equipment. Similarly, the recovery solvent must be compatible with the swab, the surface material of construction, and the analyzing test equipment. Recovery studies must be performed to document the ability of the swab to recover soil from the surface materials (typically performed with "virgin" coupons) and from the swab into the recovery solution. Finally, swabbing procedures must be developed to reduce the inherent variability in a manual process.

The drawbacks to swabbing are perceived to be outweighed by the advantage of the method—a direct, quantitative measure of surface cleanliness. When used with a product specific assay or TOC as a test method, swabbing can be an effective tool for evaluating cleaning effectiveness.

Acceptance Criteria and Limits

Why is it so difficult to establish acceptance criteria for cleaning validation studies? Part of the reason is that often the nature of the soil is not known in a biotechnology process. For example, consider the infinite number of possible combinations of protein fragments that can be produced during a cell culture or *Escherichia coli* fermentation process. Any of these protein fragments can be a con-taminating soil in the process, yet we cannot begin to know what those protein fragments actually are, let alone measure them specif-ically.

If the nature of the soil is not known, how can one determine a level of acceptability? There are several places to begin when trying to answer these questions. If one is using visual detection as an an-alytical method, then a simple question can be asked: Is any level of visible soil considered acceptable? Most answer this question with a definitive "no!" If this is the case, then the acceptance criterion for visual detection after cleaning is "none detected." If some level of vi-sual soil is to be considered acceptable, then a scientific rationale for the acceptable level must be developed and documented.

Acceptance criteria for cleaning validation must be established based on scientific rationale, and the rationale must be documented. The choice of acceptance criteria is often dependent on the product being produced or the methods used to produce the product. It should be noted that numerous assumptions are often made when applying any acceptance criteria methodology. These assumptions should be clearly understood and assessed for potential impact on the validity of the study.

Acceptance criteria may be developed based on process or clean-ing capability, percentage of toxic dose, percentage of therapeutic dose, or the limits of detection for an assay (PDA 1976). When based on process capability, cleaning acceptance criteria are set based on the ability of the process to remove the contaminant. For such crite-ria to be scientifically based, studies must be performed that show the ability of the process to remove measurable quantities of the soil of interest. Such spiking studies are commonly used for cleaning process where cross-contamination issues are a concern.

One of the most commonly used criteria for cleaning acceptance is the 1/1000th of a therapeutic dose rule. Based on existing clinical dose-response testing, this method assumes that cleaning is effec-tive as long as the contaminant is not present in a subsequent dose of product at a concentration greater than 1/1000 of a therapeutic dose (Smith 1992; PDA 1976; Fourman and Mullen 1992). Fre-quently used for multiproduct manufacturing operations, this tech-nique assumes that the soil being measured is an active compound (i.e., the cleaning process has not inactivated the compound).

LD_{50} or drug toxicity criteria are also used for situations where the contaminant has a presumed toxic effect. In these cases, the ac-ceptance criteria typically fall within the range of 0.01 percent to 10 percent of the toxic quantity (PDA 1976).

Change Control

Change control is, in many ways, as important or possibly even more important than validation itself. If validation occurs in the absence of change control, revalidation testing must be performed frequently to ensure that the system's validated state is maintained. Simply stated, once validation has been performed, the process must continue unchanged. Otherwise, revalidation must be performed.

Besides being a regulatory requirement, change control is good business practice. Maintaining records of what has changed, when, and why any changes were made helps manufacturers evaluate the positive and negative impacts any given change or collection of changes may have on the manufacturing process. Without thorough change documentation, the inevitable changes that occur in all manufacturing processes are lost. These losses, singular or cumulative, have the ultimate potential to lead a process out of control.

A good change control system will not impede change; instead, it will promote well-documented and managed change. To design a good change control program, one must recognize that change is inevitable. Change (in general) is good, and changes should be documented and evaluated for their impact on all aspects of the regulated product and manufacturing environment.

One of the major difficulties faced by change control program designers is deciding what changes should be documented. Certain changes, for instance, may be considered "critical"; other changes may be deemed "noncritical." Changing a light bulb, for example, would frequently be considered a "noncritical" change. The difficulty in making such decisions, however, is that criticality assessment is highly subjective. To one person, changing a light bulb is noncritical; however, to another, changing a light bulb is critical—e.g., if the light bulb is in a sealed light fixture in a filling area where pressure control is important. The gray area is, therefore, one of the most important challenges facing a program designer.

Efficiency is another crucial factor that affects the success or failure of a change control program. If the change control program is inefficient, one of two things is likely to occur. Either users will find ways to circumvent the program, or important changes (changes that should be made) will not be made. A change control program should not be designed to prevent or discourage change—it should complement the change process.

A typical change process flowchart is represented in Figure 7.1. A change request is initiated and then reviewed. The reviewed change request is either approved for execution or rejected and returned to the initiator. If approved, the change is executed and

Figure 7.1. Change control flowchart.

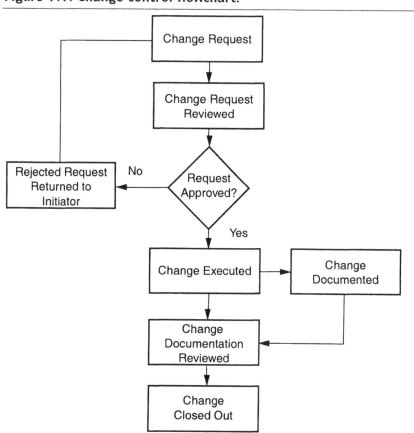

documented. Closeout of the change includes a review of the change documentation.

The two most important steps in this process are the review of the change request and the documentation of the executed change. A review of the change request must provide a mechanism whereby the effects of the change are assessed for impact on current Good Manufacturing Practice (cGMP) operations and all departments affected. Such operations and/or departments involved may include the following:

- Quality Assurance.
- Quality Control.

- Manufacturing.

- Engineering.

- Regulatory Affairs (submissions).

- Maintenance.

- Calibration.

- Documentation (SOPs, batch records).

- Training.

- Validation.

Typically, the impact assessment of the change request process is handled either by a committee or a single person. The advantage of the committee approach is that each responsible group has an opportunity to review the change and provide input to the change request; the disadvantage is that pulling together a group to review one or more change requests can take considerable time, thereby slowing the change request review/approval process. When a single person (i.e., change control coordinator) is responsible for determining the impact of a given change on all operations and/or departments, he or she becomes the critical person in the process. Such a person must be well trained to understand the ramifications of a given change. For example, the coordinator must have sufficient understanding of regulatory filings to know whether a change will affect a filing or, more importantly, must recognize when advice must be sought.

The second most important step in the change control process is documentation of the executed change. When a change request is implemented or executed, work is performed (per the approved change request) by the appropriate person or group. Documentation of the work performed is critical to the change process. In addition to recording that the work was performed, the person doing the work must document details of the actual work done and any additional work that may have been performed (often one work item leads to the need to perform more). The documentation activities are crucial for the change process to add value to the manufacturing operation.

Cleaning is so tightly woven into the fabric of the manufacturing process that change control plays a critical role in the cleaning program. Changes in the manufacturing process may impact the cleaning process and may require additional testing following a change. Changes to the cleaning cycle may be made to improve turnaround times for equipment or to accommodate different work schedules. The cleaning system itself will undoubtedly undergo change during the course of its operational lifetime. All of these types of changes must be documented and assessed for their impact on the cleaning program.

Revalidation

Revalidation (or requalification) is the term used to describe the activities that are performed in order to ensure that a validated system or process remains in a validated state—until the system is retired from service or is no longer used. Revalidation is necessary because of the inadequacy of many change control systems and because systems and equipment by the nature of their own design and use can "spontaneously" change with time. Mechanical systems, for example, experience wear as a result of continued use. If the effects of use cannot be easily detected, then requalification work is necessary to give continued assurance that the system will continue to perform as intended.

Historically, revalidation was performed primarily to evaluate change. If a major change occurred, then revalidation work should be performed. Another approach, albeit costly, was to revalidate all systems on an annual basis, regardless of need. Both approaches are inefficient and ineffective, however, and have been improved using new revalidation evaluation programs.

Many revalidation programs are now based on an annual review of three major sets of documents for each system:

- Changes to SOPs that affect the system/process.

- Change history records for the system/process.

- Repair/maintenance records for the system/process.

Assuming a thorough change control program is in effect that evaluates each change for its impact on the validation status of the system/process, revalidation can be reduced and/or better controlled. A review of the change history records for a system and the change history for appropriate SOPs, as well as maintenance/repair records, is used to evaluate the cumulative effects of change on the system/process. This is necessary because the evaluation of a single change fails to present the overall effect multiple changes may have on a system. Similarly, an individual change to an SOP may not have an effect on the way a system is operated; when changes are combined, however, they have a major impact on the way the system is operated and, hence, impact the validated state of the system/process.

CURRENT AND FUTURE TRENDS IN CLEANING

Cleaning and cleaning validation have been in the spotlight of attention, but it is now becoming a part of routine validation tasks. Cleaning validation now is at the point in its evolutionary history where the emphasis is on finding practical and pragmatic approaches to solving problems. The result of this evolution is a greater understanding of what is and what is not important, what can be done to validate cleaning, and how to do so with practical, effective tools and techniques.

People are, for instance, beginning to recognize that swabbing is just one tool in the vast array of tools available to the validation professional, and that swabbing is not the only way to verify cleaning efficacy. Similarly, TOC is recognized for what it is—a highly sensitive but nonspecific assay that is valuable as an analytical method but is not the only analytical or always the best analytical method (McArthur and Vasilevsky 1995).

Evidence of a greater understanding is supported by the trend in cleaning agent manufacturers and vendors that now will supply validation support. They will even help provide validation documentation to support the change from one cleaning agent to another.

Equipment manufacturers now have an increased understanding of the need not only to develop equipment that is cleanable but also demonstrate effective cleaning of that equipment. Sampling valves, for instance, are now becoming part of many cleaning systems. Engineers are beginning to look at equipment design from the perspective of sample collection and include ports from which samples can be easily obtained to aid in the cleaning validation program.

New cleaning techniques are being investigated, including a novel approach using solid CO_2 "pellets" to effect soil removal. Used like a traditional solvent in a high-pressure CIP spray system, the solid CO_2 "scrubs" the surface and vaporizes when finished, leaving no residual cleaning agents to contaminate future product. Advances like these, both in engineering practices and concepts will continue to improve the technologies associated with cleaning and cleaning validation.

REFERENCES

Agallaco, J. 1992. Points to consider in the validation of equipment cleaning procedures. *J. Paren. Sci. Technol.* 46 (5):163–168.

Cesar, R. A. 1994. Sanitary pumping: An update. *Pumps and Systems Magazine* (April).

Coleman, D. C., and R. W. Evans. 1990. Fundamentals of passivation and passivity in the pharmaceutical industry. *Pharm. Eng.* 10 (2):43.

Fourman, G. L., and M. V. Mullen. 1993. Determining cleaning validation acceptance limits for pharmaceutical manufacturing operations. *Biopharmaceuticals* 17 (4):54–60.

Lord, A. G. 1995. Equipment cleaning and depyrogenation. Lecture presented at the ISPE/FDA Principles of Validation and Control for Pharmaceutical Processes Course, 28 February, in Rockville, Md., USA.

McArthur, P. R., and M. Vasilevsky. 1995. Cleaning validation of biological products: A case study. *Pharm. Eng.* 15 (6):24–31.

PDA. 1976. *Cleaning and cleaning validation: A biotechnology perspective.* Bethesda, Md., USA: Parenteral Drug Association.

Smith, J. M. 1992. A modified swabbing technique for validation of detergent residues in clean-in-place systems. *Pharm. Tech.* 16:60–67.

U.S. of A., plaintiff *v. Barr Laboratories, Inc. et al.,* defendants. Civil action No. 92-1744. U.S. District Court, D. New Jersey, 5 February 1993, as amended 30 March 1993.

White, E. M., J. B. Grun, C-S. Sun, and A. F. Sito. 1991. Process validation for virus removal and inactivation. *BioPharm* 4 (5):34–39.

INDEX

Drug Manufacturing Technology Series

KEY CONCEPT CROSS–REFERENCE INDEX FOR STERILE DOSAGE FORMS

Note: This key concept index is designed to reflect all of the major topics of the volumes of the *Drug Manufacturing Technology Series* that relate to Sterile Dosage Forms as they are published. It will be expanded for each volume of the Sterile group. Volume numbers are identified by an Arabic numeral followed by a colon.

T - #0048 - 111024 - C0 - 229/152/16 - PB - 9780367400255 - Gloss Lamination